RADIO TROUBLE-SHOOTING

A Complete and Practical Work on Radio-Receiver Troubles and How to Cure Them

BY ENNO R. HAAN, E.E.

ASSOCIATE EDITOR OF
Popular Mechanics Magazine

reprinted from
an original copy loaned by:

New Wireless Pioneers
6270 Clinton St
Elma NY 14059

1928 Radio Trouble-Shooting

by Enno R. Haan, R. E.
associate editor of
Popular Mechanics Magazine

Original title: *Radio Trouble-Shooting*

Reprinted by
Lindsay Publications Inc
Bradley IL 60915

ISBN 1-55918-010-2

1 2 3 4 5 6 7 8 9 0

1989

PREFACE

Radio receivers, like all other electrical and mechanical devices, are subject to wear and depreciation, and a certain amount of trouble is bound to arise sooner or later, even in the best of instruments. The correction of radio trouble is a subject that is touched upon only by occasional magazine articles and small booklets, and very little material of this nature has been published in book form. At the time of this writing there has not yet appeared a complete book entirely devoted to radio trouble shooting, and it is therefore the author's aim to meet the widespread demand for dependable and authoritative information on this subject.

While simple enough for the veriest amateur, this book is at the same time complete and exhaustive, and will appeal specially to the service man. The subject is treated in a thoroughly practical manner, with concrete instructions and numerous illustrations showing the reader how to test parts of a receiver, how to trace trouble and remedy it, and how to take care of the receiver properly in order to avoid trouble. The contents of this book are based on actual practice, and theory on the subject is eliminated except so far as is necessary to enable the owner to locate and remedy trouble.

Indebtedness is hereby gratefully acknowledged to *Popular Mechanics Magazine,* from which journal some of the photos and illustrations used in this book were obtained, and also to Mr. J. B. Smith, E.E., of the Radio Service Co., Chicago, for suggestions and assistance in insuring the accuracy of the material of this book.

THE AUTHOR.

December 5, 1927.

TABLE OF CONTENTS

TABLE OF CONTENTS

CHAPTER I

TOOLS AND INSTRUMENTS

Every radio owner should have a few simple tools and instruments for testing, maintaining, and repairing his receiver. Proper tools for this purpose will facilitate the work considerably, and will pay for themselves in a short time by the saving on repair bills. However, it is not advisable for the average fan to purchase the more expensive instruments, which he would seldom use, unless he intends to go into radio service work as a spare-time occupation.

Headphone Tester.—Fig. 1 shows a headphone tester, which is generally useful for the detection of open circuits, short circuits,

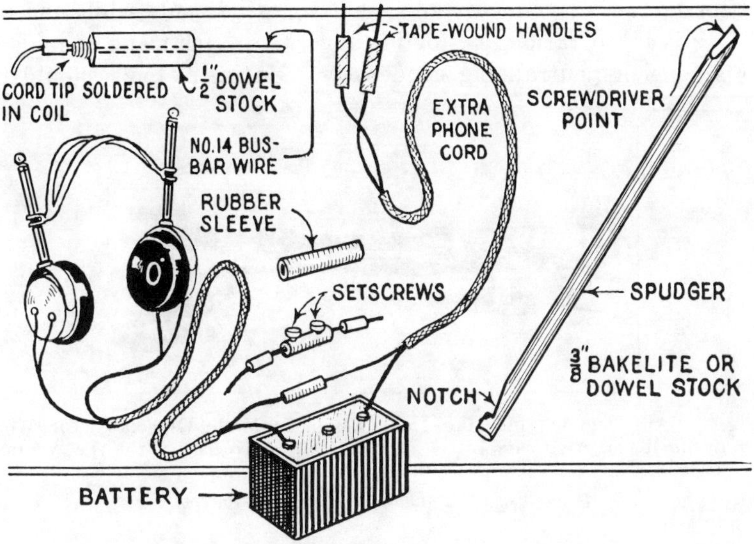

Fig. 1. The Headphone Tester Is the Most Universal Instrument for Making Tests on Faulty Receivers and Instruments, and a "Spudger" Is also Indispensable for the Radio Trouble-Shooter

grounds, and faulty conditions in the receiver and in the accessories. It consists of a battery connected in series with an ordinary pair of headphones, and two specially made testing tips. It is provided with an extra cord, as indicated, to facilitate connecting and disconnecting the battery. The use of an extra phone cord is handier than separate flexible leads. A connector of the type shown in the drawing will be found handy for connecting cord tips together, a short length of rubber tubing being slipped over it so that there will be no possibility of short circuits by accidentally getting it in contact with the battery terminals. Testing tips with insulated handles should be provided. They are made as follows: Get a length of ½-in. dowel stock and cut off two pieces, each about 2 in. long. Drill a hole through the center of these pieces, the hole being just large enough to permit a length of No. 14 round copper wire to be pushed through. The end of the wire is bent to form a small coil in which the cord tip is soldered. Rubber tape is then wound over the wooden handles to provide good insulation.

Spudger.—Adjustments must sometimes be made while the receiver is in operation, as, for instance, the adjustment of midget balancing or neutralizing condensers. No metal tools should be

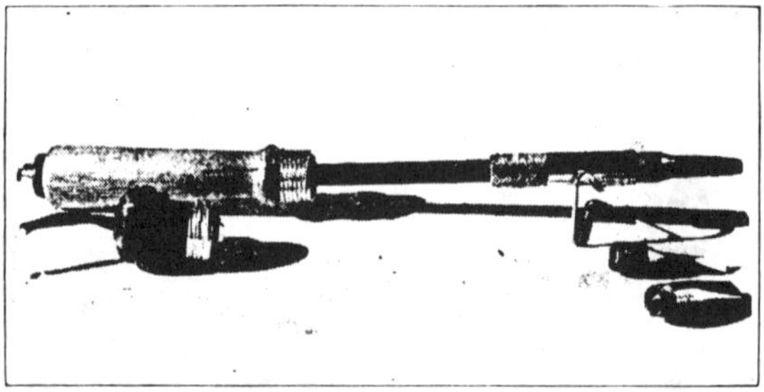

Fig. 2. An Electric Soldering Iron, Which Is Made Especially for Work on Radio Receivers, Costs Little and Is Really a Necessity for the Average Radio Owner

(Courtesy of "Popular Mechanics Magazine")

used for this work, as the capacity of the operator's body and of the tool may interfere with the accuracy of the adjustment. It is better to use a "spudger," which is also shown in Fig. 1. This simple tool consists of a 6 or 8-in. length of ⅜-in. wooden or bakelite rod. One end is ground or filed down to a screwdriver edge, and the other end has a notch cut in it. Besides being useful for making adjustments this tool will also be found handy for locating loose connections and for cleaning and bending up socket prongs, the notch being used for the latter purpose.

The spudger can be inserted between the wires without any danger of short circuits, but care must be taken not to bend the wires, as this may cause them to come in contact with each other and cause a short circuit.

Soldering Iron and How to Use It.—In radio receivers it is absolutely essential that all the connections are mechanically and electrically tight. If this is not the case, there may be considerable loss of energy due to high-resistance joints, or reception may be noisy due to loose connections. For this reason all

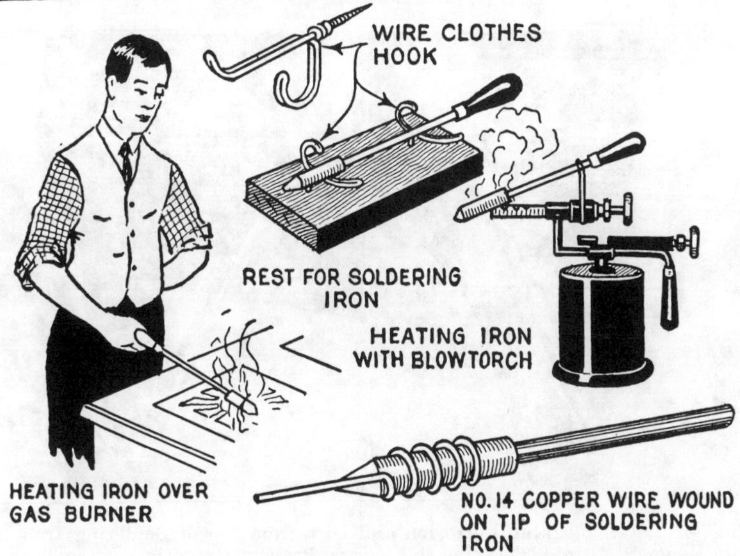

Fig. 3. Heating a Soldering Iron over a Gas Burner or Blow Torch, a Rest Which Prevents Burning the Bench Top, and a Length of Copper Wire Wound on Tip of Iron to Solder in Narrow Places

wires attached to soldering lugs must be securely soldered. A cheap electric soldering iron, which can be obtained for about a dollar, will meet the needs of the average fan, who uses it only occasionally. A better grade of electric iron is advisable for the service man as he uses it considerably. Such an iron, having removable tips of different sizes, is shown in Fig. 2. A soldering iron used for radio purposes must have a long, narrow nose to enable the user to reach into places which are inaccessible with a heavy iron of the kind used by tinners. When it is necessary to solder joints that cannot even be reached with a small iron, wrap a 12-in. length of heavy copper wire around the nose of the iron and let part of the wire extend as shown in Fig 3. The wire will then be heated sufficiently to melt solder.

Some sort of rest is necessary to lay the hot iron on while it is not in use. A rest for this purpose can easily be made by driving two wire clothes hooks, of the usual type, into a small piece of 1-in. wood, which is about 6 in. long and 4 in. wide. After the hooks are driven into the wood, the projecting screw ends can be filed down, and the hooks are then bent over to a position

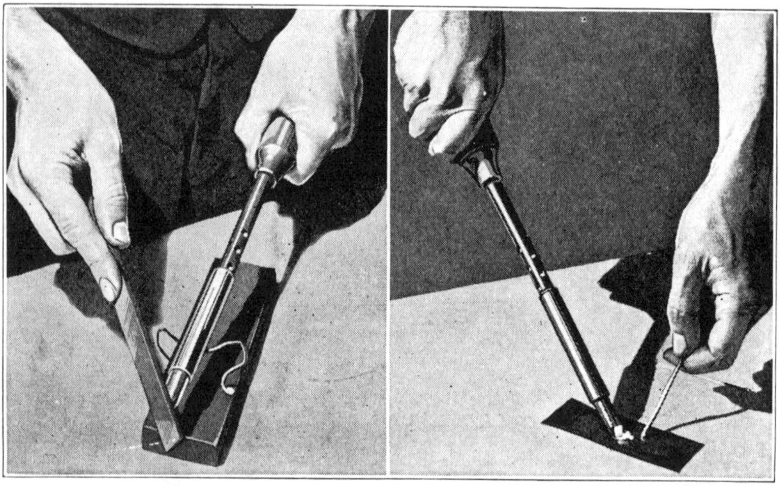

Fig. 4. Left: Cleaning Corrosion and Dirt from Tip of Soldering Iron Before Using It. Right: Tinning the Iron to Prevent Oxidation of the Tip and to Facilitate the Conduction of Heat to the Work

(Courtesy of "Popular Mechanics Magazine")

most convenient for holding the iron. Such a rest will prevent burning the top of the work bench by laying the hot soldering iron on it.

Where electricity is not available, the use of an electric soldering iron is out of the question, and an ordinary soldering iron must be used. A small one can be obtained for about 25 cents. If gas is available, the iron can be heated by laying it on the burner over the flame. A tinner's blow torch must be used if there is no gas supply. When using these methods of heating a soldering iron, be careful not to let the iron get too hot. Never let it get red hot, as this will cause a film of copper oxide to form over it and then it will be necessary to file this off before the iron can be tinned so as to be fit for further use.

The working tip of a soldering iron should always be kept clean and well "tinned," as shown in Fig. 4. Brighten the tip with a knife or with an old file and rub the hot iron over a piece of tin, copper, or brass to which flux and solder have been applied, as also shown in Fig. 4. The tip of the iron will then have a thin film of solder over it, which is required to do good work. Retinning is frequently necessary while the iron is being used. Corrosion and dirt should never be permitted to accumulate on

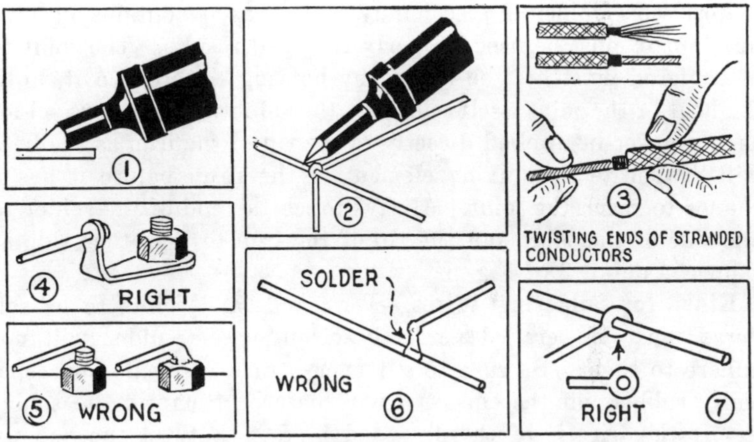

Fig. 5. The Correct and Incorrect Methods of Using a Soldering Iron, How to Connect Wires to Each Other and to Terminal Screws, and How to Twist the Ends of Stranded Conductors

the tip of the soldering iron, as this will result in poorly soldered and dirty joints. After the iron is sufficiently hot and well tinned, the flat side of the tip is held in contact with the joint to be soldered, as shown in detail 1 of Fig. 5, and a little flux is applied. After the joint has been thoroughly heated, the solder is applied sparingly. Do not use too much solder, as large, bulky joints will then result; it is better to use just enough solder to flow in between the joint and a little over it. When the iron is removed the solder will harden in a few seconds, and the joint should then be wiped to eliminate all excessive flux and dirt.

Solder and Flux.—The purpose of flux is to clean the joint and permit the solder to adhere to it. Ordinary solid wire solder can also be used, but this requires the application of flux. A paste or a rosin-solution flux will be found convenient and non-injurious to parts of the receiver. Never use an acid flux in a radio receiver under any circumstances, for it will gradually eat away the metal parts including the wires, and will cause considerable corrosion, even long after the soldering has been completed. In some cases wires have been completely eaten away by acid flux. A good flux to use, which can be applied with a small camel's-hair brush, is a solution of powdered rosin and alcohol.

Rosin-core solder is also handy to use, as it contains its own flux, but it must be used properly for good results. The joint to be soldered must first be heated by holding the iron on it, until the heat of the joint itself will melt the solder. Rosin-core solder should never be applied directly to the tip of the iron as the heat will evaporate the fluxing element in the rosin before it has a chance to reach the joint. Always touch the end of the solder to one side of the joint and the tip of the iron to the other side as shown in detail 2 of Fig. 5.

Kinks for Soldering Wires.—Wire ends and joints to be soldered must be scraped perfectly clean or the solder will not adhere to them. Be sure to get every trace of insulation, especially rubber, off the end of a wire before attempting to solder it. If the wire is of the stranded kind, first twist the end up tight, as shown in detail 3 of Fig. 5. When connecting wires to the screw terminals of instruments it is the best practice to use soldering lugs, as shown in detail 4 of Fig. 5, instead of twisting the end of the wire around the screw or soldering them directly to the end of the screw, as shown in detail 5. Besides providing a much neater appearance, soldering lugs make a better contact, as they are flat, and the nuts, when screwed down tightly against the lugs, are not so apt to come loose. The correct and incorrect methods of joining two wires together are shown in details 6 and 7 of Fig. 5. Never butt the end of one wire against the side of another, but make a loop on the end of the former, hook it over the other wire, and then pinch the loop tightly around the latter wire with a pair of small pliers, after which it can be soldered.

Miscellaneous Tools.—Fig. 6 shows a number of tools which

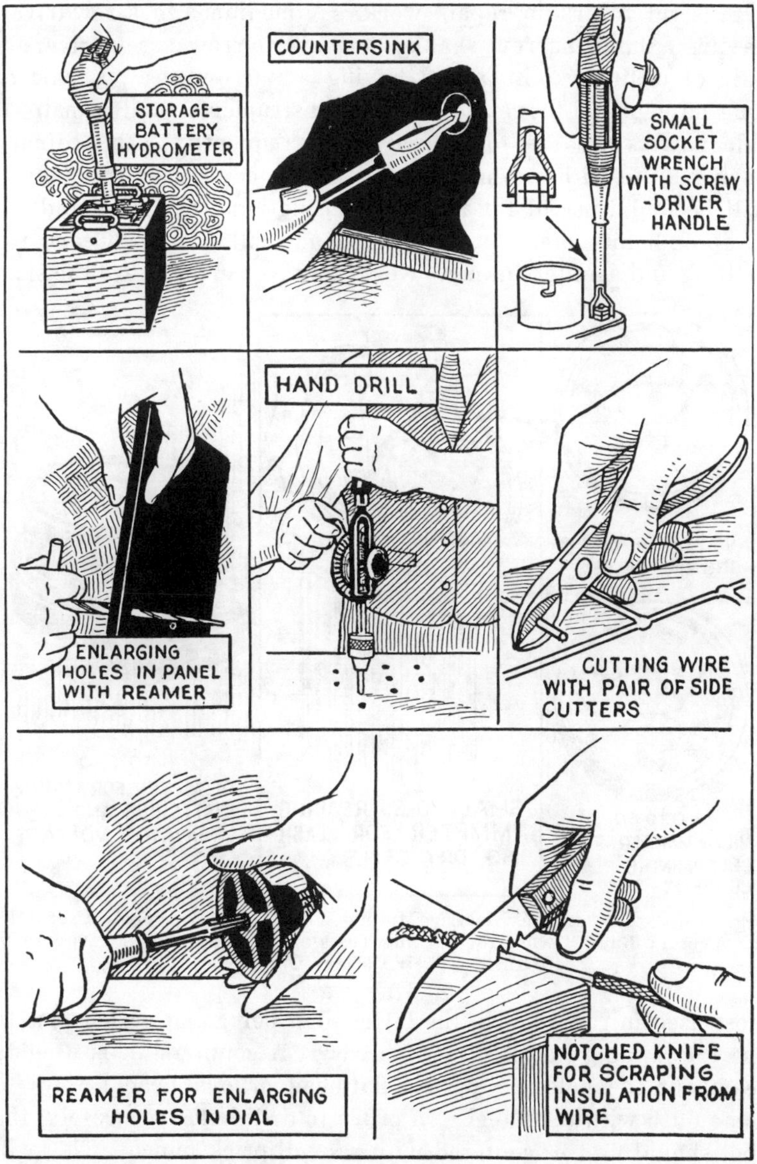

Fig. 6. A Number of Tools Which Will Be Found Useful by the Radio Trouble-Shooter. A Hydrometer, a Hand Drill, Wire Cutters, Socket Wrench, and a Knife Are the Most Important

are useful for radio repair work. A medium-size screwdriver, having a long, narrow shank; a small, narrow screwdriver; a pair of ordinary 5-in. pliers having a narrow nose; a pair of long-nose pliers to get in between instruments, and a pair of side cutters are the tools used most frequently. An ordinary paring knife with two notches filed in the edge, near the handle, will be found useful for removing insulation from wire, and for scraping joints, etc. An ordinary hand drill is often needed for drilling and countersinking. No. 27 and No. 36 drills are the sizes

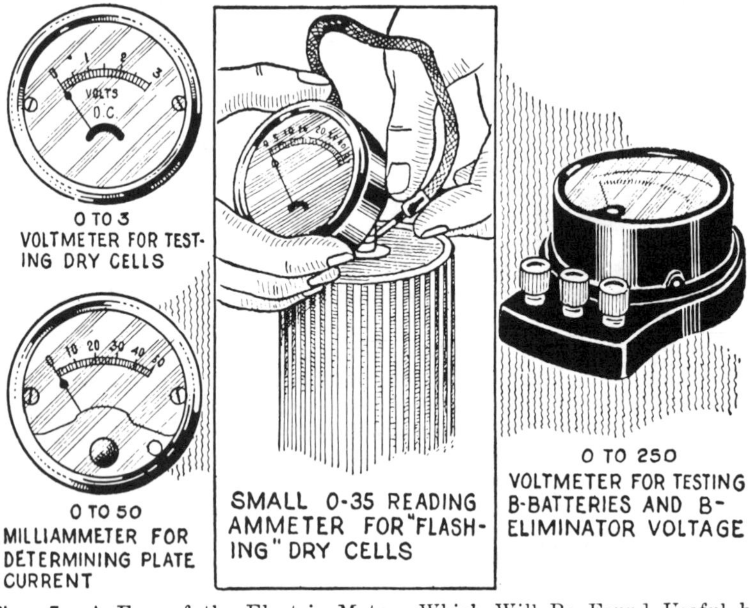

Fig. 7. A Few of the Electric Meters Which Will Be Found Useful by the Average Radio Owner for Making Tests on His Receiver and Apparatus to Insure Their Correct Operation

most used in radio work, the latter drill being the proper size to use for tapping holes for 6–32 screws A countersink is used to ream out the holes to permit flat-head or oval-head screws to come flush with the panel. In order to drill holes accurately, the spot should first be centered with a small prick punch. This will prevent the drill from shifting before it seats itself. A small flat file and a small round file are often useful. Occasionally there is use for a hacksaw. Reamers to enlarge holes in the panels and dials are often convenient. A couple of screwdriver-type socket wrenches, to fit the small and the large nuts of 6–32 screws will be found useful to every radio owner. If the owner has a storage battery, a hydrometer is necessary to indicate the condition of charge. The use of this instrument is thoroughly explained in the section on storage batteries. A strip of emery cloth, some fine sandpaper, friction tape, some soldering lugs, a few lengths of No. 14 copper, bus-bar wire, and a few lengths of "spaghetti" tubing will also be found handy.

Meter Equipment.—Some of the most useful meters for radio purposes are shown in Fig. 7. The condition of B-batteries and the voltages applied to the plate circuit of a receiver can be determined accurately only by means of a voltmeter. When a

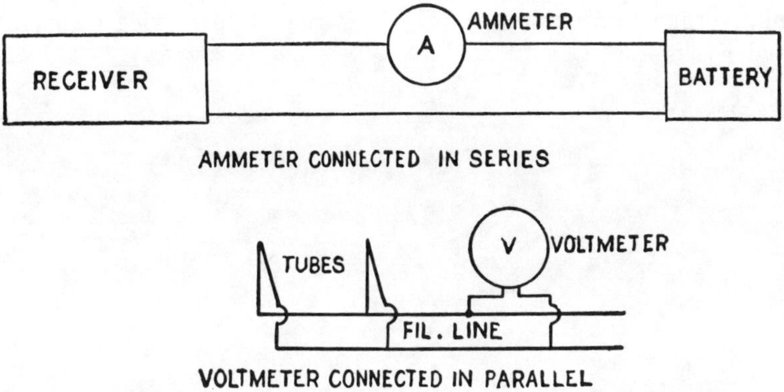

Fig. 8. In Electrical Circuits an Ammeter Is Usually Connected in Series with the Current-Consuming Device, While a Voltmeter Is Customarily Connected in Parallel

B-eliminator is used to supply plate current, a high-resistance voltmeter is necessary to determine the exact output. A low-resistance voltmeter cannot be used for this purpose because it consumes so much current that it gives an erratic reading during the test. The voltmeter used for reading plate voltages should cover a range of from zero to the highest voltage obtainable from the B-batteries or the B-eliminator used.

A small pocket-type ammeter, having a range of from 0 to 35 amperes, is generally used for "flashing" dry cells. However, a more accurate method of testing dry cells is to use a voltmeter, reading from 0 to 3, and graduated in tenths of a volt. A milliammeter is not essential for the average fan, but will often be found useful to the service man. If it is to be used for testing the plate current of the individual tubes, it should have a 0 to 10-milliampere range, and if used for testing the total plate-current draw of a receiver, it should have a 0 to 50 range. It should be remembered when using voltmeters and ammeters, that, as a rule, the former are connected directly across the terminals of a device of which the voltage reading is to be taken. Ammeters, on the other hand, are hooked up in series with the device or devices of which the current consumption is to be ascertained. Fig. 8 shows the proper method of using voltmeters and ammeters.

Tube testers, tube reactivators, and wavemeters are described in other sections. These instruments are usually not needed by the average fan, but are required by the radio service man and dealer.

CHAPTER II

UNCONTROLLABLE TROUBLES AND INTERFERENCE

The troubles to which radio receivers are subject are many and varied; some of them are caused by faulty or worn-out equipment, such as batteries and tubes, or defects in the receiver and loudspeaker. However, there are certain troubles to which all receivers are subject, and these, as a rule, cannot be eliminated by any efforts on the part of the radio owner. Some of these are sporadic and intermittent, while others are permanent. To this class of troubles belong static, fading, and total inaudibility due to "dead spots." Poor or defective broadcasting equipment also causes trouble to owners of radio receivers, and this, in a way, is beyond their control. There are other interference troubles, some of which can be eliminated by the radio owners. Such troubles are power-line hum, crackling noises caused by sparking brushes on motors, and interference caused by instruments that can be shut off while the receiver is being used.

Atmospheric and Geographical Conditions

Static.—A lightning discharge in the immediate vicinity of a radio receiver produces a loud crash in the loudspeaker. More distant discharges produce irregular, frying sounds. The varia-

Fig. 9. The Effect of Static on Radio Waves Can Be Likened to Huge Waves Intercepting a Beam of Light. However, Static Discharges Can Also Be Heard in Radio Receivers

(Courtesy of "Popular Mechanics Magazine")

tion of charge and the dissipation of electricity from the clouds also causes crackling and hissing noises. This is static. Sometimes it seems as if the air is entirely devoid of signals, and it seems as if an impenetrable blanket is thrown over the vicinity, although no abnormal noises are evident in the receiver. Static then intercepts radio signals and prevents them from reaching the receiver. This may be likened to the interception of a light beam by high waves, as shown in Fig. 9. Such conditions may be local, existing in a small territory of a few square miles, or they may cover a vast area of several thousand square miles. Sometimes they exist for only a few hours, and at other times they last for several days. It has been found that static is usually more troublesome during the summer months than during the winter months, and it is also more evident in tropical and semi-tropical regions than in regions farther north.

Receivers having long aerials pick up more static disturbances than receivers provided with short aerials, indoor aerials, or loops. There is really no method of eliminating static at present, although some devices on the market are supposed to do

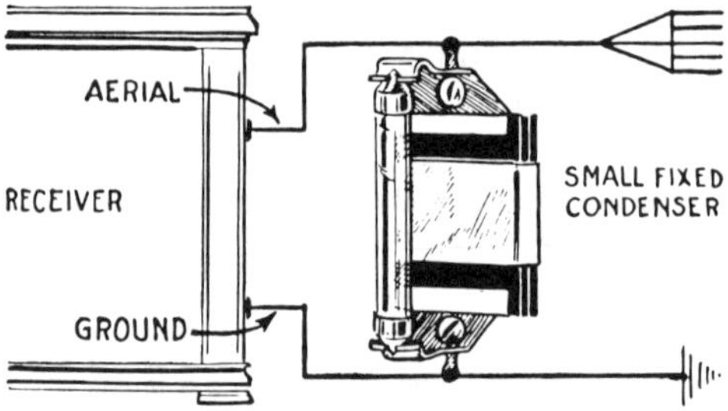

Fig. 10. Small Condenser and Leak Between Aerial and Ground Reduces the Effect of Violent Static Crashes

this. However, the effects of static in a receiver can be reduced to some extent by shortening the aerial and connecting a small-capacity condenser in series with the aerial or ground lead, as described in the section entitled, "Influence of Aerial Height and Length." It is also a good idea to use a fixed resistor across the condenser, as shown in Fig. 10, which will tend to by-pass severe static discharges. Loose coupling of the tuning device of a receiver will help to reduce the static pick-up, but loose coupling also results in the weakening of station signals, which is, of course, undesirable.

Fading.—One may have noticed, when a distant station is tuned in on a receiver, that the signal varies in volume, sometimes becoming almost or entirely inaudible, but finally coming back again to its original volume. This is called "fading," and it is caused by varying atmospheric conditions between the receiver and the broadcasting stations. Sometimes the signal strength of the station varies, and this is mistaken for fading. When fading is noticed, do not attempt to increase the filament current of the tubes beyond their proper operating point, nor attempt to retune the dials, for the signal may then be lost entirely.

"Dead Spots."—Certain spots on the earth's surface are apparently impervious to radio signals, and these are called "dead

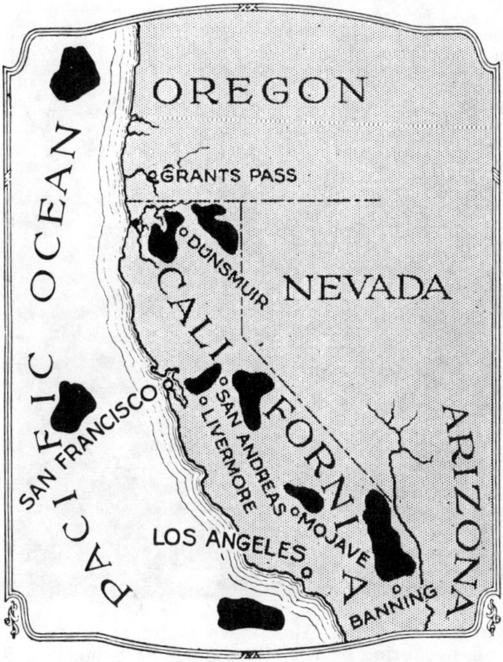

Fig. 11. Map Showing the "Dead" Spots in California. Radio Can neither Be Sent nor Received in These Areas
(Courtesy of "Popular Mechanics Magazine")

spots." About fifty of these have been charted in North America, and they have usually been found accidentally, so that it is probable that there are many more. One of the largest dead spots is located a little south of the Hudson Bay in Canada; another is over the ocean off Atlantic City; while a third is in the vicinity of Camden, N. J. There are numerous dead spots in Mexico and the southwestern states. Fig. 11 shows a map giving the approximate location of dead spots in California. Radio signals can be received all around these spots, but the dead areas cannot be penetrated, even when receivers are placed close to their boundaries. The cause of these phenomenal conditions has not been determined. They exist on both land and sea.

Shielding Effects of Steel Structures.—Sometimes it is impossible, in a certain location, to get reception on a receiver which has been found to work elsewhere, and in such cases it is possible that the particular location of the receiver is shielded

Fig. 12. In Large Office and Apartment Buildings Having Considerable Steel Framework Reception of Distant Signals Is Sometimes Impossible on Loop Receivers

from external electrical impulses, and hence the receiver cannot pick up radio signals. This is usually the case when a loop-operated set is installed in a steel-constructed building, the kind shown in Fig. 12. The steel absorbs most of the energy and grounds it. This trouble and its remedy are thoroughly described in the section entitled "Loop Troubles." Similar blanketing effects are often evidenced when receiving apparatus is located in the vicinity of large deposits of metal in the earth.

Faulty Broadcasting Equipment and Operation.—The conversion of sound waves to electrical impulses, which are radiated from the aerial of a broadcasting station, is a delicate task. Fig. 13 shows the interior of a typical broadcasting station and some of the apparatus required. The instruments used for doing this are subject to limitations, even when handled by experts. Often poor instruments, especially a microphone, which is also shown in Fig. 13, produce distortion. Faulty adjustment of the transmitting instruments, as for instance the tuning equipment, will also cause trouble for radio owners, especially those in the

Fig. 13. Left: Some of the Apparatus Used for Radio Transmission in a Large Broadcasting Station. Right: A Microphone, Which Picks up the Sounds and Changes Them into Electrical Impulses

immediate vicinity. Broadcasting stations are licensed to operate only at certain wavelengths, and when radio owners can prove that a station is not adhering to its allotted wavelength, complaints may be made to the government radio supervisor of the district.

Interference Troubles

There are many sources of radio interference over which the radio owner has some control. In a few cases he may be able to remedy the trouble at its source, and in other cases he may prevent interference from entering his receiver.

Reradiation Interference.—One of the most frequent interference troubles is that experienced from reradiation of regenerative receivers. All receivers, which feed back some of the plate current from one or more tubes into the aerial, either directly, as is the case with the "ultra-audion" hookup, or indirectly, as is the case with three-circuit tuners, are regenerative. When these receivers are operated improperly, the tubes oscillate and the aerial is charged with a high-frequency current, which is radiated in the air for many miles, and is picked up and amplified by all the receivers in the vicinity, as shown in Fig. 14. This interference causes either a steady howl or a whistle; or, if the tuning dial on the offending receiver is being rotated, "mockingbird" whistles of varying intensity will be heard in other re-

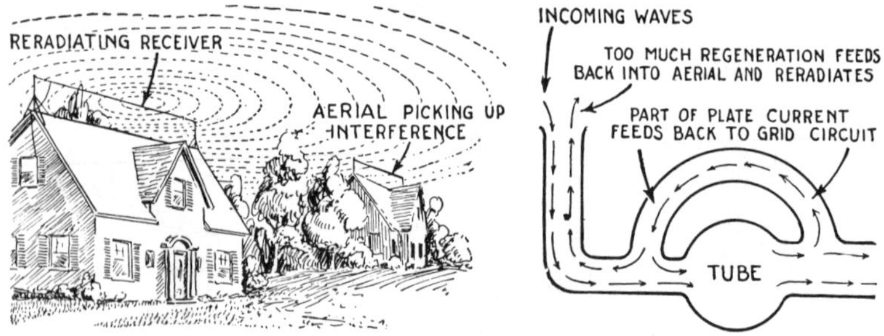

Fig. 14. Left: Electro-magnetic Waves Are Radiated from the Aerial of a Regenerative Receiver. Right: Analogy Showing how Excessive Feed-Back Current Returns into the Aerial

ceivers. The trouble cannot be eliminated at the receiving end, although it can be reduced to some extent by methods that also decrease the signal strength of stations tuned in, and this is, of course, not desirable. The trouble can usually be eliminated at its source. As a rule, it is caused by turning up the tubes too high. The use of a non-reradiating unit on the offending receiver is an effective cure. Such a unit is described in a later section entitled "Non-Reradiating Unit."

Interference from Transmitting Sets.—Telegraph transmitting sets also are a frequent source of annoyance to radio owners. This kind of interference is usually noticeable at the lower wavelengths, and depends to some extent on the proximity of the transmitter. Amateur transmission is allowed only on wavelengths below 200 meters, but occasionally this waveband is overstepped, and signals are transmitted in the broadcasting waveband, which is from 200 to 600 meters. Amateurs are deprived of their license to operate and their transmitting sets are sometimes confiscated after repeated offenses. Complaints should be made to the government radio supervisor of the district. Proximity of a receiver to a transmitting station also accounts for considerable interference, even if the transmitter is tuned below 200 meters. Rotary spark gaps used in transmitting sets cause a constant crackling noise, which somewhat resembles static disturbances, and is often mistaken for the latter. This interference will continue even though no signals are sent, and it can often be heard all over the dials of nearby receivers.

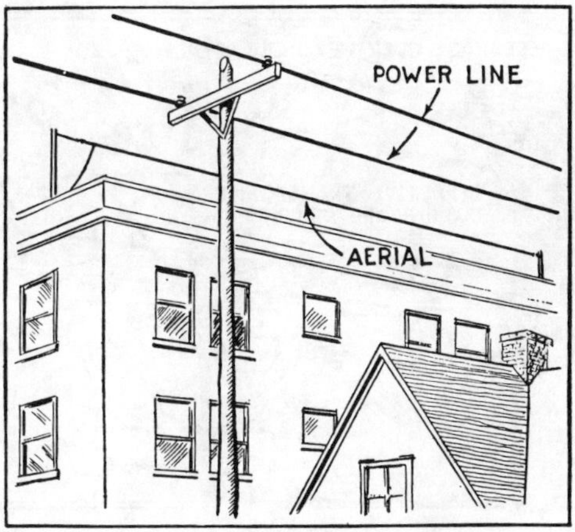

Fig. 15. Aerials Should never Be Erected Parallel to Electric Power Lines

Power-Line Hum.—Power lines are frequently a source of interference. Aerials erected parallel to power lines, as shown in Fig. 15, intercept the electro-magnetic field, which exists around a power line, and this induces current in the aerial at the same frequency as that in the power lines. The current thus induced in the aerial is transferred to the receiver and is amplified, producing a steady and annoying hum in the loudspeaker. This trouble and its cure are explained in the section entitled "Aerial Parallel to Power Lines." Obviously, loop receivers are not subject to this trouble as much as aerial receivers, but nevertheless, if the loop or the receiver is placed in close proximity to electric-lighting wires or extension cords on lamps, the same trouble may result. Such interfering currents can also be picked up by the coils in the receiver. Shielding the coils often helps to eliminate the trouble entirely. The method of completely shielding a receiver against this and other interference is described in the section entitled "Complete Shielding of the Receiver." Inductive effects similar to those caused by power lines may also be caused by telephone and telegraph lines, and by street-car and interurban-trolley lines. In such cases, however, the effect in the loudspeaker may be different; the noise may be more irregular, depending on the time when these lines are used. Loud buzzing and crackling sounds may then be evident.

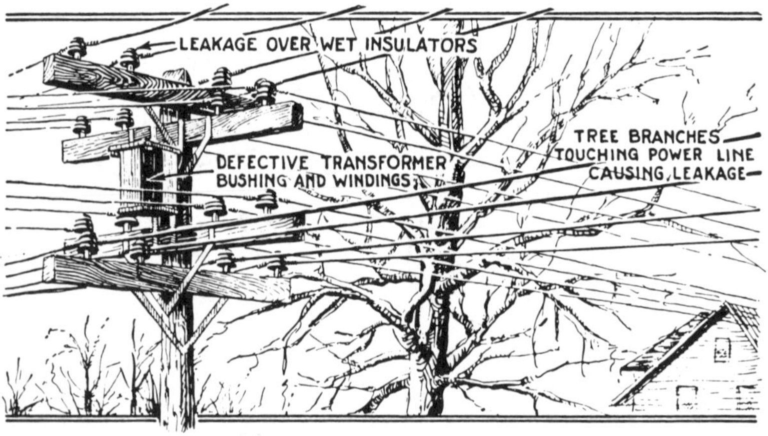

Fig. 16. Leakage of Electricity over Wet Insulators, over Tree Branches, and Defective Power-Line Transformers May Cause Crackling Noises in Radio Receivers

Grounded Power and Telephone Lines.—In case a power line is partly grounded, crackling noises will be heard in the receivers in the entire vicinity. Grounded conditions are frequently caused by the branches of trees touching the power lines, as shown in Fig. 16. Whenever this happens, there may be a leakage of current at this point, accompanied by sparking or arcing. As a rule such grounds are intermittent, due to the branches swaying in the wind and touching the power line occasionally, and for this reason, the source of the trouble is often difficult to locate. It is a good idea to go around the vicinity to look for such tree grounds, while the trouble is being experienced. A large percentage of external interference evidenced by crackling sounds in the receiver is caused in this way. Intermittent grounds of a similar nature can also be caused by wet insulators; see Fig. 16. This usually results in a steady buzz in the receivers. Faulty transformers are sometimes to blame for grounded conditions, but as a rule these are kept in good condition by the power companies.

Interference from Power-Line Devices.—Dirty commutators and sparking brushes on electric motors used on washing machines, vacuum cleaners, fans, oil-burning furnaces, refrigerators, sewing machines, etc., are a frequent cause of crackling

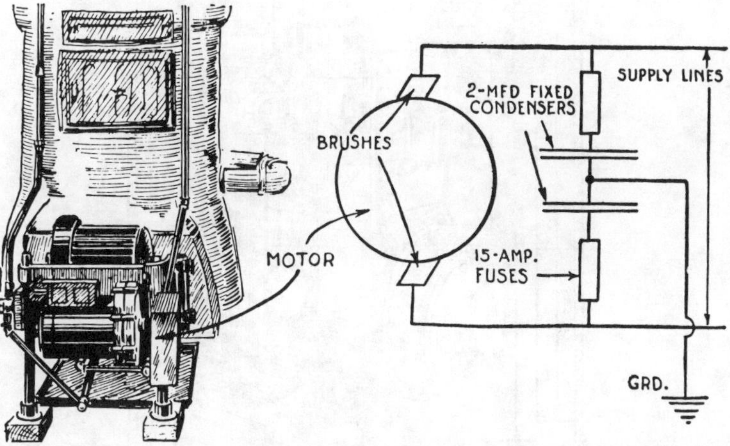

Fig. 17. Oil-Burning Furnaces, Which Have Electric Motors, often Cause Considerable Crackling Noise in Radio Receivers, Resulting from Dirty Commutators and Sparking Brushes

noises, if these devices are used while the radio set is in operation. In some cases where these motors cannot be turned off, as is the case with those used on oil-burning furnaces, the trouble can often be eliminated as shown in Fig. 17. Connect two 2-mfd. fixed condensers in series across the power line near the motor which is causing the trouble. A ground connection on a cold-water pipe is then made to a point between the two condensers. Fuses of 15-amp. capacity are also provided between the line and the condensers as indicated, to prevent short circuits in case the condensers break down. The operation of high-frequency devices, such as violet-ray and X-ray outfits, ozonators, induction coils, and the like, causes continuous crackling or buzzing sounds. Other sources of interference from power-line devices are: sign flashers, electrically operated switchboards used for railway signaling, elevator controllers, automatic cutouts and starting switches for machinery, stop lights on street corners, street arc lamps, motion-picture machines using arc lamps, electric-welding apparatus, smoke or dust precipitators, toy electric motors, toy electric trains, doorbells and buzzers.

Interference from Farm-Lighting Plants.—Farm-lighting plants are equipped with a magneto, which sometimes causes consider-

Fig. 18. By Shielding the Farm-Lighting Plant, Interference Trouble from This Source Can Be Eliminated

(Courtesy of "Popular Mechanics Magazine")

able interference. The high-tension lead from the magneto to the spark plug should be shielded by slipping a length of flexible metal tubing over it. The ends of the tubing should not be allowed to touch the lugs on the wire, and therefore it is a good idea to wind tape around both the tubing and the leads. An insulated wire is then soldered to the tubing and connected to a water pipe or to a metal rod driven into the ground. Use a ground clamp to get a good connection, if the metal is such that it cannot be soldered. Similarly, shield the magneto by means of a metal can or box if the magneto is accessible, taking care, however, to mount the shield in such a way that it will not touch any part of the magneto. Then connect this shield to the metal tubing on the high-tension lead, soldering the wire to both. Use insulated wire, which will eliminate all danger of short circuits. Another method of shielding the plant is shown in Fig. 18. In this case the whole plant is enclosed in a sheet-metal box, which is grounded.

It is also advisable to prevent interference caused by sparking commutator brushes. The method of doing this is also indicated in Fig. 18. Get two condensers from discarded Ford coils, being sure, however, that the condensers are in good condition. This can be determined by testing them according to the instructions contained in the section entitled "Testing Condensers." The two condensers are connected in series across the output lines of the generator, and a wire is then connected to the center point between them and to the ground direct, or to the magneto shielding, which is grounded. It is a good idea to insert 15-amp. fuses between the output lines of the generator and the condensers to prevent a short circuit in case one or both of the condensers break down. A further precaution is to provide inductance coils in the line, as shown in the diagram. These coils are wound on a tube 2 in. in diameter and 3 in. long, and consist of approximately 150 turns, the same kind and size of wire being used as that used for the output lines of the generator. Obviously, farm-light plant interference is evidenced only while the plant is in operation, and not while it is standing idle. If the owner does not desire to go to the trouble of shielding and by-passing the currents causing the interference, as just described, the trouble can be eliminated by operating the light plant only at times when the receiver is not in use.

Other Electrical Devices Causing Interference.—When using automatic telephones, a clicking will be heard in the receiver while the calling dial of the telephone is being rotated. This trouble is, however, temporary. Street cars, interurbans, and elevated trains produce considerable sparking on the trolleys and rails, especially if the latter are dirty, and this sets up interference waves which make radio reception almost impossible. The ignition systems of automobiles produce high-frequency currents, which also interfere with reception. Some common causes of interference are shown in Fig. 19.

How to Locate Interference.—One can readily determine whether interference evidenced by crackling and spluttering sounds is caused by some power device or loose connection in the electric-lighting circuit of the building in which the receiver is located, by simply throwing out the main switch of the circuit, if such is provided. In case there is no switch, loosen the fuses

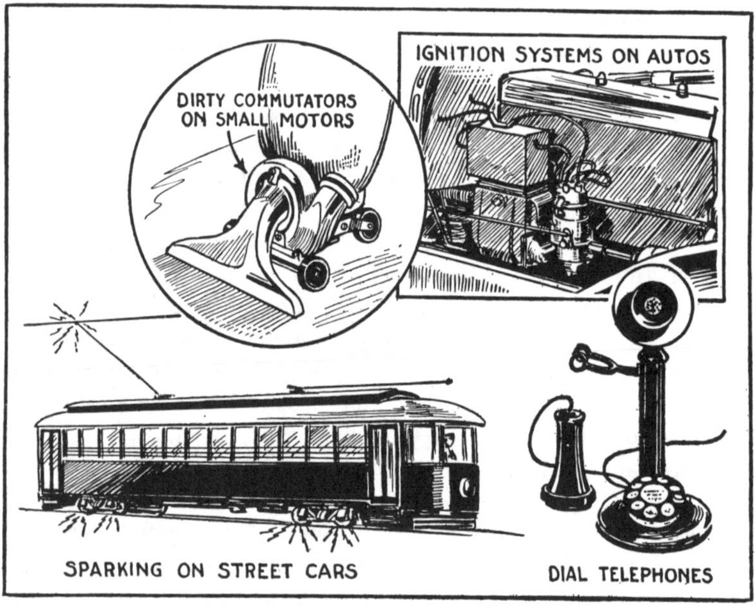

Fig. 19. Some of the Causes of Interference in Receivers. Dirty Commutators on Motors Are Frequent Offenders

one by one, so that no current can flow through the circuit. Then switch on the receiver, and if the interference has ceased, one may be reasonably certain that the trouble lies in the circuit. The next step is to find the trouble. Throw the supply switch in, or screw the fuses tight. Then, while the receiver is operating and the interfering noise is heard, turn off all motors and other electrical appliances which are in operation. Also switch off the circuits of the individual rooms, and tighten all cord plugs which fit in flush receptacles or lamp sockets. If there is any lamp that flickers, there may be an internal loose connection in the socket, or the lamp may not be screwed into the socket securely, resulting in a loose contact.

If the interference does not stop when the main switch is thrown out, or the fuses are removed, then the source of the trouble does not lie in the lighting circuit. The trouble can be located by using a portable receiver operating on a loop aerial. The whole vicinity should be canvassed, the receiver being set in operation at various spots, and the loop being rotated toward the point where the interfering signal comes in loudest. Carefully note in which direction the loop points when the greatest strength of the interfering signal is observed. By making a complete circle it is usually an easy matter to locate the source of the trouble.

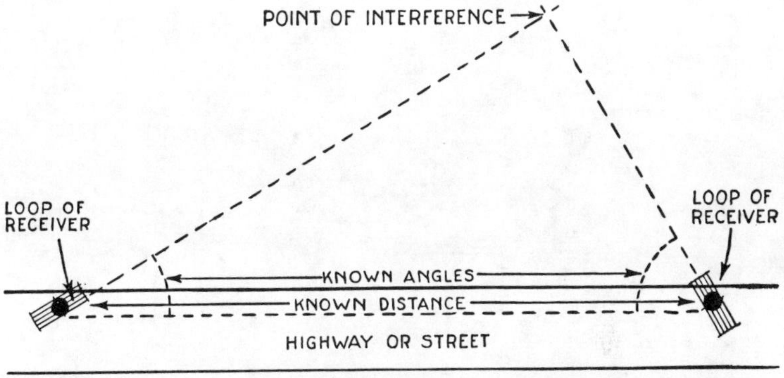

Fig. 20. An Unknown Point from Which Interfering Signals Are Radiated Can Be Located by the Use of a Loop Receiver Tuned to Bring in the Interfering Signal at Two or More Locations

If the point of interference is quite a distance away, and the task of encircling it would take a good deal of time, it can be approximated by the method shown in Fig. 20. The loop receiver is tuned to pick up the interfering signal at two points along the street or highway, as indicated, and the loop is turned toward the direction from which the trouble seems to come. In both cases the angle of the loop in relation to the direction of the street or highway must be carefully noted. The distance between the two points where the interfering signals are tuned in must also be known. A drawing similar to the one shown here is then made, a straight line being drawn to represent the distance between the two points where the interfering signal was tuned in, and the angles are then laid out. Two dotted lines are extended along the sides of the angles as shown to represent the direction in which the loop was pointed. The intersection point of the dotted lines represents the place from which the interference is radiated. It is then an easy matter to measure the dotted lines and determine what distance they represent, comparing them with the line of which the length is known. According to the distances thus obtained, the approximate point of the interference can be calculated.

CHAPTER III

ANTENNA-CIRCUIT TROUBLES

Every radio receiver must have some method of picking up the electrical impulses sent through the air by broadcasting stations, and this is done by means of an antenna and a ground, or a loop antenna alone. In colloquial radio parlance, an antenna is more commonly referred to as an aerial. Without an aerial and ground, or a loop, radio reception is impossible at present, although there are receivers that have a small pickup coil or rod inside of the cabinet; but such devices, too, are essentially aerials. An aerial picks up the radio-frequency current and conducts it through the tuning coil of a receiver to the ground, through which it makes its return to the broadcasting station, as shown in Fig. 21. If there is an open circuit or a grounded condition in the aerial system, the path for the current through the receiver is broken, and it is then usually impossible to obtain

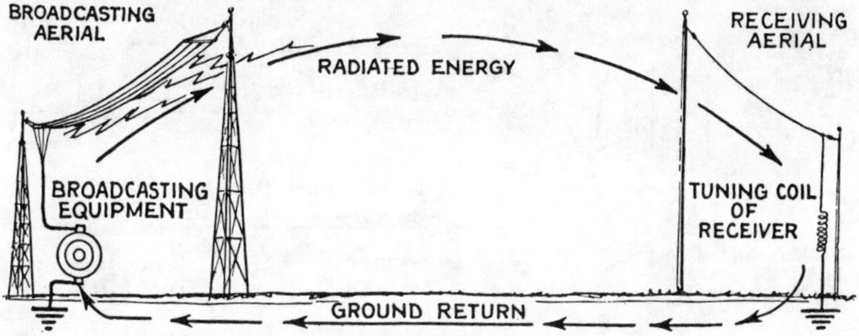

Fig. 21. Path of Radio Current Leaving a Broadcasting Station, Passing Through the Tuning Coil of the Receiver and Returning to the Source of Supply. Loop Receivers Work Differently—See Fig. 51

reception. Often such trouble is caused by broken, loose and corroded connections, which may be unnoticed unless a careful inspection is made. When installing an aerial there are many points to bear in mind, as for instance, the length, height, kind of wire used, its relation to other aerials and power lines, the mechanical strength of the installation, and its electrical condition. These are all factors which greatly affect reception, and trouble may result if they are not taken into consideration when installing the aerial.

How to Install Outdoor Aerials

Erecting the Aerial.—In spite of the simplicity of doing the work, poor aerial and ground installations are quite common. Carelessness is usually the reason for this, although many radio owners are not aware of the importance of a good installation, such as is shown in Fig. 22. When an aerial is to be erected, first make observations as to where it will be most convenient. A high aerial is advisable, and if possible, it should be kept away from other aerials, or at least installed at right angles to them. Look around for electric-light and telephone lines and be sure not to run the aerial parallel to any of these, for troubles from interference may then result.

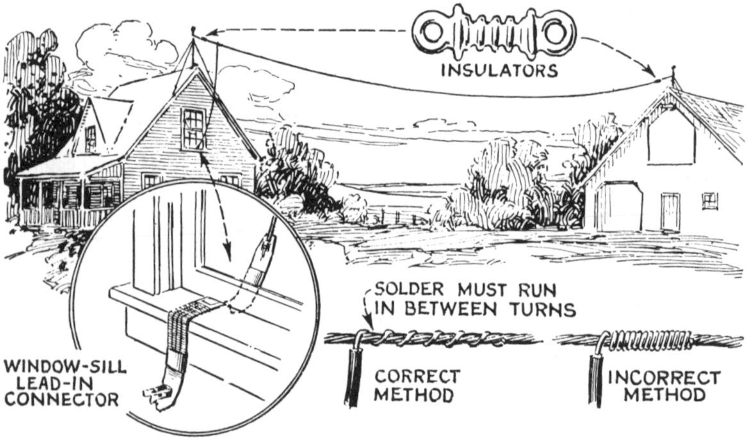

Fig. 22. Good Aerial Installation, Showing Where Insulators Are Used, How the Lead-in Wire Is Brought into the House and the Correct and Incorrect Methods of Attaching the Lead-in to the Aerial

After a good location for the aerial has been found, and suitable supports or masts are provided, tie a length of stranded wire to one support, letting one end of the wire project for the attachment of an insulator. The wire used for the aerial should not be smaller than No. 22. Stranded, hard-drawn, copper or phosphor-bronze wire is preferable, and it is connected to an insulator at each end. Be sure to pull the aerial up tightly, for it will sag to some extent, and allowance should be made for this. The aerial should be at least 3 ft. above the roof and higher if possible.

The Lead-in Wire.—The lead-in wire, which should be No. 14 stranded wire having weatherproof insulation, should be bared at one end for about 6 in. The bared section is scraped clean with a knife, and is twisted around the aerial at the end closest to the receiver. Allow plenty of space between the separate turns of wire; never twist them together tightly, as it will then

Fig. 23. Soldering the Lead-in Wire to the Aerial Is Necessary to Obtain the Most Efficient Operation of a Radio Receiver
(Courtesy of "Popular Mechanics Magazine")

be difficult to flow the solder in between the turns and make a good electrical connection. Soldering flux is first daubed on the joint, which is then heated with a soldering iron or blow torch, after which the solder is applied, as shown in Fig. 23. The soldering iron is held on the connection until the solder runs between all the turns. It is absolutely important to make a soldered joint here if one wishes to obtain good reception.

Run the lead-in wire down to the receiver. It must be held by insulators, as shown in Fig. 24, at points where it is apt to rub against projections on the building, such as cornices and gutters, for a grounded condition will result if the insulation of the lead-in wire is rubbed off and the bare wire makes contact with such projections. Usually it is most convenient to run the lead-in wire through a window. When this is done, it is advisable to provide a flat window-sill connector, as shown in the insert of Fig. 22, so that the window can be closed on it. Another

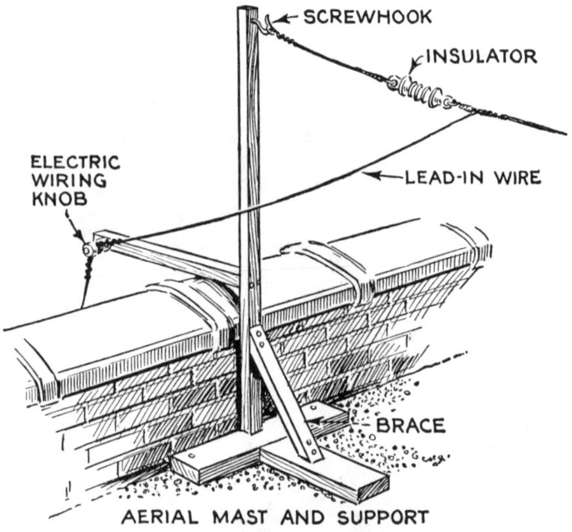

Fig. 24. Aerial Mast on a Flat Roof Showing Method of Attaching the Aerial and Lead-in Wire, the Latter Being Held by Means of a Porcelain Knob to Prevent Contact with the Coping

advantage of using such a connector is that it eliminates the danger of fraying off the insulation from the lead-in wire by closing the sash on it, and the resulting danger of breaking the lead-in wire at this point.

Installing a Lightning Arrestor.—Although almost every radio article in magazines and newspapers dealing with aerial installations stresses the importance of a lightning arrestor, and the same advice is usually given in the manufacturers' instruction sheets accompanying most receivers, many owners neglect to install this apparatus. Approved lightning arrestors are so cheap and easily installed that there is no excuse for not having one. Besides, the arrestor will not only protect the receiver in case of a lightning discharge, but it may also aid in obtaining payment of insurance, if a fire occurs. The usual error made when connecting a lightning arrestor is shown in Fig. 25, where the arrestor is connected in series with the aerial and the receiver. Such an installation makes it impossible to obtain reception, and the trouble is often blamed to the lightning arrestor. The correct method of installing it is also shown.

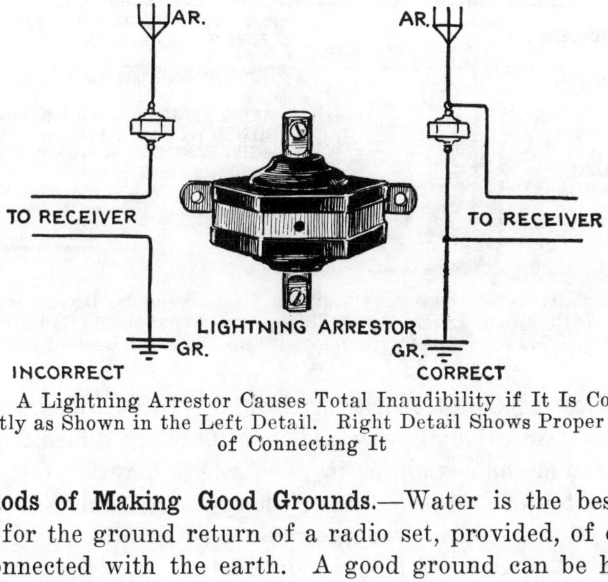

Fig. 25. A Lightning Arrestor Causes Total Inaudibility if It Is Connected Incorrectly as Shown in the Left Detail. Right Detail Shows Proper Method of Connecting It

Methods of Making Good Grounds.—Water is the best conductor for the ground return of a radio set, provided, of course, it is connected with the earth. A good ground can be had by making a connection with the water in a well or cistern, with a cold-water pipe, or with a metal rod or pipe driven down into moist earth. A metal rod driven into dry surface soil makes a poor ground connection. Various methods of providing efficient ground connections are shown in Fig. 26. When a ground connection is made to a radiator, to a water pipe, or to the conduit of an electric-wiring installation, a separate ground connection should be provided outside of the building for the lightning arrestor. It is absolutely necessary, when using clamps to make a ground attachment to a water pipe, to brighten the pipe by scraping or filing at the point where the clamp is to be attached, for if there is a film of corrosion, rust, or paint on the pipe there will be a high-resistance joint, which will materially decrease the efficiency of a receiver. A poor ground connection may even prevent reception entirely.

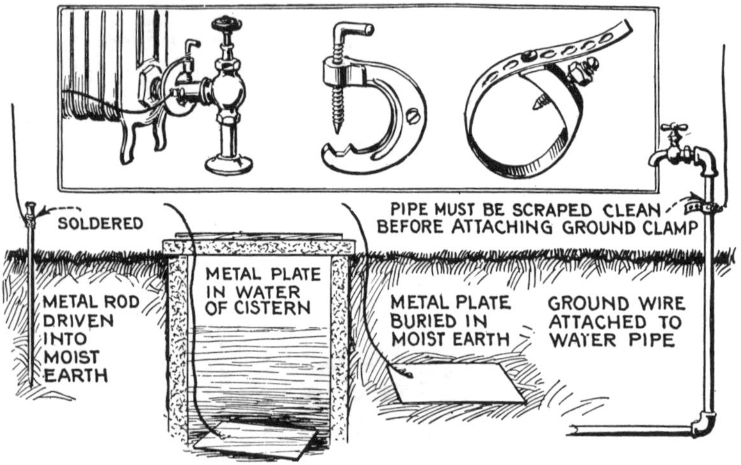

Fig. 26. Various Methods of Obtaining Good Grounds, Direct or Indirect Contact with Moist Earth Being Necessary. Above: A Radiator Which Usually Makes a Good Ground and Two Types of Ground Clamps

Substituting Counterpoise for Ground.—In locations where the ground is exceptionally dry and it is therefore difficult to get a good ground connection, or in cases where a receiver is located near the top of a high building, where the ground wire must be of exceptional length, a counterpoise may be substituted for the ground. A counterpoise is merely another wire just as long as the aerial and erected in exactly the same way; it is located below the aerial and parallel to it, as shown in Fig. 27.

Material Used for Aerials.—As an aerial is constantly exposed to the weather, it should be sufficiently strong to stand up against such exposure. Also, the wire used for the aerial must have maximum conductivity. Hard-drawn copper or phosphor-bronze wire has been found satisfactory; it can be obtained at any shop handling radio supplies. Besides providing greater surface area, which increases conductivity, stranded wire is much stronger than solid wire, and it is flexible, so that it can be bent without breaking.

Ribbon aerials, which consist of a solid copper or phosphor-bronze ribbon, are claimed to have a higher pick-up value than ordinary stranded wire, but according to the U. S. Bureau of Standards, little advantage is gained by the use of ribbon aerials, as the additional pick-up value is so slight that it is practically negligible. Furthermore, the use of ribbon aerials is more or less objectionable, as sleet is likely to freeze on them and the increased weight may cause them to break. Ribbon aerials also offer considerable resistance to the wind, and when severely strained are likely to break off near the insulators. Hollow, woven-wire aerials also have the objectionable features of ribbon aerials, and are therefore not recommended. Solid-wire aerials may be used in emergencies, when stranded wire is not available,

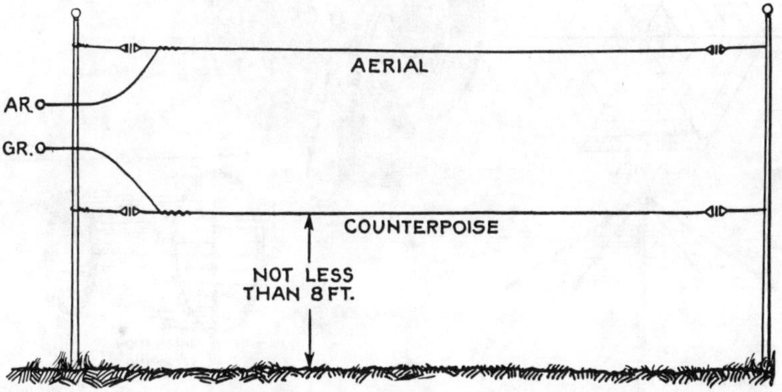

Fig. 27. In Locations Where the Installation of a Good Ground Is Impossible, a Counterpoise of the Same Length as the Aerial and Erected Directly Under It as Indicated, Should Be Used

but this wire gets brittle and is likely to break near the insulators, as a result of constant swaying. Instead of using bare, solid copper wire when regular aerial wire is not obtainable, it is better to use insulated wire, which is used by electricians for house-wiring purposes, as this is much stronger than bare wire. Insulation on wire does not appreciably reduce its pick-up quality, as radio currents penetrate walls and everything they come in contact with.

Substitutes for Regular Wire Aerials.—Many pick-up devices, which are essentially aerials, have been placed on the market. Some of these are in the form of hollow spheres, wire-wound poles or plates, and wire cages of various shapes and sizes. Some of these aerials are shown in Fig. 28. They absorb some of the radio energy in the air, which is then conducted to the receiver over the lead-in wire. The chief merit of these devices is that they are handy to install, and on congested apartment buildings in large cities, a number of these aerials do not present such an objectionable appearance as a maze of wires radiating in all directions. Tests indicate that these devices have less pick-up value than a regular wire aerial and therefore they will be found sharper for local reception.

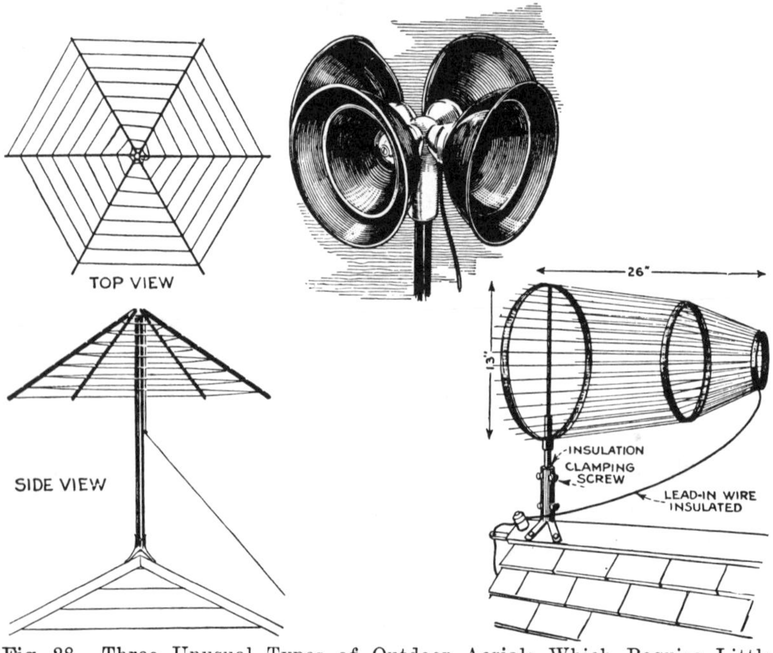

Fig. 28. Three Unusual Types of Outdoor Aerials Which Require Little Space, Are Entirely Non-Directional, and Are Easy to Install
(Courtesy of "Popular Mechanics Magazine")

If the radio owner in a city wishes to install one of these devices, but does not feel justified in going to the expense of purchasing one, he can make one himself from two sheets of copper or brass, about 1 ft. square. A slit is cut in each sheet, diagonally from one corner to the center of the plate as shown in Fig. 29. The sheets are slipped over each other as indicated and soldered together in this position. A binding post is also soldered to the edge of one sheet, and the device, thus completed, is suspended from two insulators attached to a wooden frame. However, in locations where sharpness of tuning is not highly necessary, and maximum pick-up value is desired, such a device offers no advantages, and it is advisable to use a regular wire aerial.

Single-Wire versus Multi-Wire Aerials.—For all practical purposes the radio fan will find that single-wire aerials will give just as good results as aerials having several conductors, con-

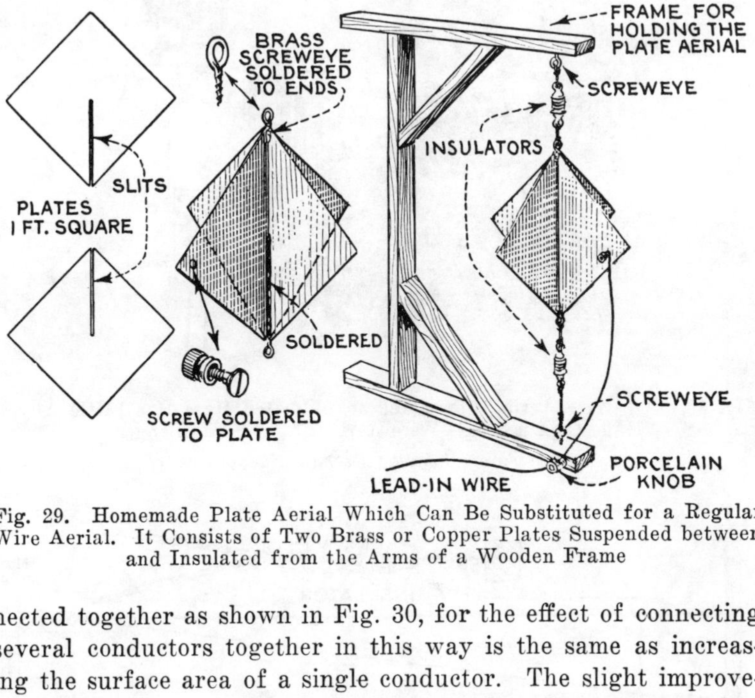

Fig. 29. Homemade Plate Aerial Which Can Be Substituted for a Regular Wire Aerial. It Consists of Two Brass or Copper Plates Suspended between and Insulated from the Arms of a Wooden Frame

nected together as shown in Fig. 30, for the effect of connecting several conductors together in this way is the same as increasing the surface area of a single conductor. The slight improvement that such a multi-wire aerial may give is not appreciable. A single-wire aerial is much easier to install, costs much less, and is therefore preferable to a multi-wire aerial.

"Zig-Zag" Installation of Aerial.—In many locations, especially in cities where it is impossible to erect an aerial in one straight line, a "zig-zag" installation, as shown in Fig. 31, may be necessary. A number of insulators and screwhooks are obtained. The length of the aerial which is best suited to the receiver must first be determined accurately, and then its exact position. The screwhooks are located at various points as indi-

Fig. 30. A 3-Wire Aerial, Connected as Indicated Here Has Little Advantage Over a Single-Wire Aerial of the Same Length

(Courtesy of "Popular Mechanics Magazine")

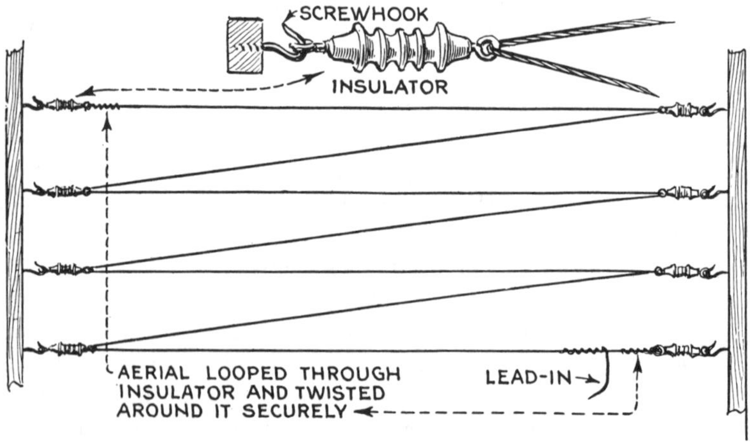

Fig. 31. Where the Space for Erecting an Aerial Is Limited to Less Than 30 Ft., It Can Be Strung Back and Forth from Several Insulators Held on Screwhooks

cated, the insulators are hooked on, and the aerial is strung through, being securely tied to the insulators at both ends. The lead-in wire is soldered on in the usual way.

Influence of Aerial Height and Length.—There are several factors that determine the best aerial length and height for any particular receiver. Often specific instructions in this respect are furnished by the manufacturer of the receiver, but as he can only advise according to average conditions, the radio owner should be able to determine to some extent what aerial length is best suited to his particular needs. It makes a great difference whether he lives in a congested city and is surrounded by a dozen or more broadcasting stations, or whether he is in the

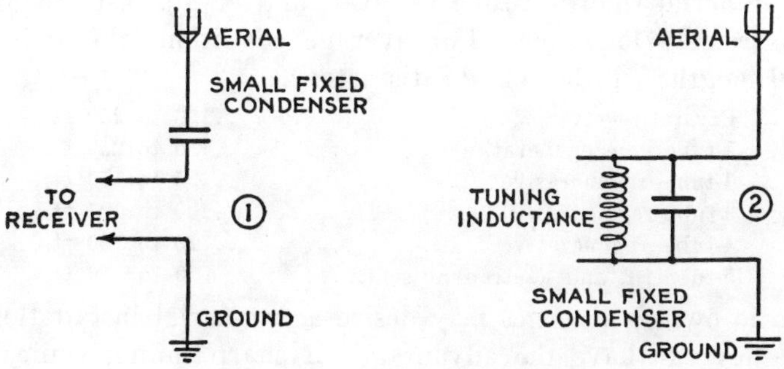

Fig. 32. (1) Considerable Selectivity Can Usually Be Gained by Connecting a Small Fixed Condenser in the Aerial Circuit. (2) Shunting a Small Fixed Condenser Across the Tuning Coil of a Receiver Reduces the Tendency to Oscillate

open country, a few hundred miles from the nearest broadcasting station. In the former case he needs extreme selectivity, perhaps at a sacrifice of volume and long range, while in the latter case the requirements are just the opposite. For selectivity, a low, short aerial will be found to give best results. If, after shortening the aerial, the receiver still tunes too broad, it may be necessary to connect a small fixed condenser in series with the aerial or ground lead, as shown in detail 1 of Fig. 32. The capacity most effective for any particular receiver varies with the capacity of the tuning inductance, but such values as .00025 and .0005 mfd. are often used for this purpose. It may be necessary to do some experimenting with various sizes in order to find a condenser which is most effective.

If an aerial is too short, there will be a lack of pick-up value, which is evidenced by weak reception. Sometimes a short aerial causes a receiver to go into oscillation easily. A method of reducing this trouble is to shunt a .00025-mfd. fixed condenser across the terminals of the tuning inductance, as shown in Fig. 32, detail 2. For distance and volume, it will be found that a long aerial will give the best results. However, a long aerial has the disadvantage of picking up more static and electrical disturbances than a short aerial. Extremely broad tuning can often be attributed to an excessively long aerial.

The length of aerial most suitable for the average receiver depends to a great extent on the number of tubes it has and the tuning-circuit design, a 5-tube receiver usually employing a shorter aerial than a 1-tube receiver, and crystal sets requiring an extremely long one. For average conditions the following aerial lengths will be found satisfactory.

Crystal receiver	150 to 250 ft.
1-tube non-regenerative	150 to 250 ft.
1-tube regenerative	125 to 150 ft.
3-tube regenerative	100 to 125 ft.
4-tube regenerative	75 to 100 ft.
5-tube r.f. and neutrodyne sets	60 to 80 ft.

Radio owners in cities may desire a two-aerial installation, so that they can have the advantages of sharp tuning while local

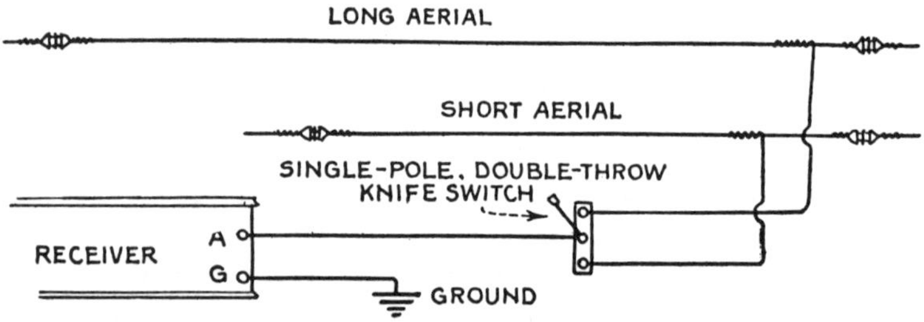

Fig. 33. Where the Advantages of Both Long Distance and Selectivity Are Desired, a Long and a Short Aerial, Connected to a Single-Pole, Double-Throw Knife Switch, May Be Used

stations are operating, and also of long range when the local stations are silent, for in most large cities a weekly "silent night" is observed to give local fans a chance to bring in some distant stations. A two-wire aerial installation is shown in Fig. 33, both aerials being provided with separate lead-in wires, which are connected to the end terminals of a single-pole, double-throw knife switch. The switch enables the operator to use either the long or the short aerial at will.

Directional Qualities of Aerials.—It has been found that aerials manifest a slight directional tendency, which is more pronounced the longer and the lower the aerial is erected. The pick-up value is greatest for broadcasting stations directly in line with the aerial, and toward which the lead-in end of the aerial points. If one desires to use this tendency of aerials to advantage in reception, two or more aerials, pointing in different

Fig. 34. The Directional Quality of Outdoor Aerials Is Made Use of in This Installation

(Courtesy of "Popular Mechanics Magazine")

directions, should be erected, as shown in Fig. 34. If only two aerials are used in this way, the lead-in wires are connected to a single-pole, double-throw knife switch. If more than two aerials are used, the lead-in wires can be connected to switchpoints as shown in the detail.

Two-in-One Aerial Installation.—In some cases it may be found convenient to install two aerials end to end, with an insulator between them, as shown in Fig. 35. Such an installation is especially adapted for use where two radio owners find it difficult to erect separate aerials. The lead-in wires are attached to the extreme ends, as indicated. It is advisable to erect the aerials in such a way that they can readily be lowered in order to clean the insulator between them. Soot and moisture may permit leakage from one aerial to the other. No inductive relation, however, will exist between them.

Fig. 35. Two Aerials Can Be Erected End to End with an Insulator Between Them, the Lead-in Wires Being Attached to the Extreme Ends. There Is No Mutual Induction from Aerials Arranged in This Way

Concealed Lead-in and Ground-Wire Installation.—When making a permanent installation of an aerial and ground, it will be found that the method shown in Fig. 36 is not only neat, but as the lead-in wire is entirely concealed, there will be less danger of experiencing trouble with it. Drill a 5/8-in. hole in the siding near the roof of the house where one end of the aerial is attached. Lower a long length of ordinary furnace chain through the hole so that it hangs inside of the wall, keeping hold of the end so that the chain will not drop down into the partition entirely. When the chain is shaken from the position at the top it can be heard inside of the house. A small hole is made in the wall between the studs where the chain is heard, for the location of a flush receptacle or a special jack to which both lead-in and ground wires are to be connected. Then, with a short length of wire having the end bent to form a hook, the chain is "fished" out through the hole in the wall, the bared end of the lead-in wire is attached to it as indicated, and the chain with the lead-in wire attached to it at the bottom is then pulled up. The same procedure is followed to install the ground wire, lowering the chain from the hole inside of the house and "sounding" it from the outside of the house before drilling the hole through the siding just above the foundation. Short porcelain tubes are slipped over the lead-in and ground wires at the points where they pass through the siding. Fasten the lightning arrestor on the outside of the building, drilling holes for the wires that connect to the lead-in and ground wires. All the wires running to the receptacle or jack must be long enough so that they can be attached before the receptacle or jack is fastened in place. A long iron rod or pipe driven about 4 ft. into the ground will usually make a good ground connection.

A plug to fit in the receptacle or jack is obtained and the lead-in and ground wires from the receiver are connected to its two terminals. These two wires should be of different colors so that

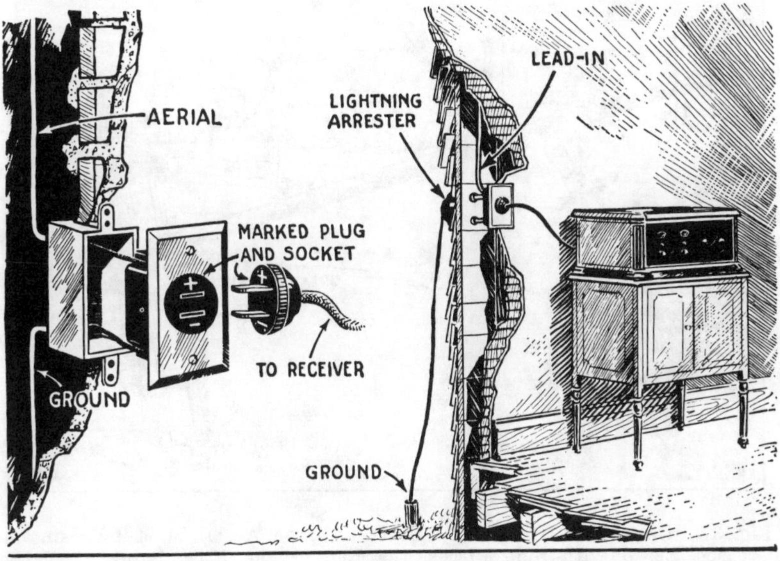

Fig. 36. Concealed Installation of the Aerial Lead-in Wire, Dropped down in a Partition, an Ordinary Plug and Socket Being Used for the Aerial and Ground Connections

they can readily be distinguished, and the terminals to which they are connected should also be marked so that the aerial and ground connections will not be confused. A good method of doing this is to put a daub of red paint on one terminal of the plug, and another on the corresponding clip of the receptacle. By following the drawing carefully, anyone can easily make such an installation.

Erecting Aerial Masts.—To slide a long aerial mast into the hole dug in the ground is a rather difficult task. However, anyone can do this without any trouble by following the method shown in Fig. 37. A board about 6 in. wide is first set up in the hole, strips of wood having been previously nailed to the edges to prevent the mast from slipping off. If the mast is made of pipe, a wooden plug is driven into the end so that it will slide down the board more easily. The plug should be wide and flat on the end to prevent the pipe from turning, and the edges should be rounded off. Both the plug and the board should be given a liberal coating of cup grease or ordinary lard to reduce friction.

Fig. 37. One Man Can Readily Erect a Large Aerial Mast by Using a Greased Board with Side Pieces, the End of the Mast Being Similarly Greased to Reduce Friction

Bracing Aerial Mast on Edge of Roof.—It is sometimes necessary to erect aerial masts near the edge of a roof to obtain the maximum distance between them for the aerial. When this is done, it is a rather difficult task to provide a brace which will take the strain of the aerial. A good method of bracing such an aerial mast is shown in Fig. 38. A strong guy wire is attached to the top of the mast, slipped through a screweye driven into a horizontal length of wood, which extends out over the edge of the roof, and is then attached to the side of the building as indicated. Screweyes and screwhooks will be found convenient for making these attachments. If a turnbuckle is available, it can be used as indicated, which makes it possible to tighten up the guy wire occasionally.

Fig. 38. Good Method of Bracing an Aerial Mast on a Flat Roof to Compensate for Aerial Strain

Underground Aerial Reduces Static Interference.—A specially designed underground aerial, called the "Subantenna," will, it is claimed, eliminate much interference and static by using ground waves instead of the air waves, because the former are believed to be practically free from atmospheric disturbances. This type of antenna consists of a shielded and sheathed wire having specially constructed pick-up coils at the ends and at the center. It is available in either 75 or 100-ft. lengths. The installation of this device is simple. Fig. 39 illustrates how this is done. Choose a suitable location under the window nearest the radio set, and dig a hole about 3 ft. deep or more, and 2½ or 3 ft. in diameter. Starting with the knob end of the aerial, lay it on the bottom of the hole, winding it in a clockwise direction toward the center, to form a flat, spiral coil. Keep the turns about 1 in. apart, until the space left in the center is about 6 in. across. Cover the coil thus formed with a 3-in. layer of earth, carrying the wire in a long sweep through this earth to the side of the hole again; then lay another coil and cover it in the same way, continuing until four coils are made. A 75-ft. length will leave a part to be used as a lead-in wire. Five or six coils can be wound with the 100-ft. aerial. If the basement floor is not cemented, the installation can be made in the basement. There are two other methods of installing this type of aerial: in a straight-line trench from 18 in. to 2 ft. deep, or in a trench of the same depth, but in the shape of the letter "S." The method explained above and illustrated is best, as the spiral form is non-directional.

Fig. 39. Method of Installing an Underground Aerial, Which Is Claimed to Pick Up Less Static and Other Interference
(Courtesy of "Popular Mechanics Magazine")

Underwriters' Rules Regarding Aerial Installations

Aerial installations for radio-receiving sets should be in accordance with the following rules as established by the National Board of Fire Underwriters, and they are given here so that the radio fan may acquaint himself with them.

For Receiving Stations Only—

a) Antenna and counterpoise outside of buildings shall be kept well away from all electric-light or power wires of any circuit of more than 600 volts, and from railway, trolley, or feeder wires, so as to avoid the possibility of contact between the antenna or counterpoise and such wires under accidental conditions.

b) Antenna and counterpoise, where placed in proximity to electric-light or power wires of less than 600 volts, or signal wires, shall be constructed and installed in a strong and durable manner, and shall be so located and provided with suitable clearances as to prevent accidental contact with such wires by sagging or swinging.

c) Splices and joints in the antenna span shall be soldered, unless made with approved splicing devices.

d) The preceding paragraphs, **a)**, **b)**, and **c)**, shall not apply to light and power circuits used as receiving antenna, but the devices used to connect the light and power wires to radio-receiving sets shall be of approved type.

e) Lead-in conductors shall be of copper, approved copper-clad steel, or other metal which will not corrode excessively, and in no case shall they be smaller than No. 14, except that bronze or copper-clad steel not less than No. 17 may be used.

f) Lead-in conductors on the outside of buildings shall not come nearer than 4 inches to electric-light and power wires, unless separated therefrom by a continuous and firmly fixed non-conductor which will maintain permanent separation. The non-conductor shall be in addition to any insulating covering on the wire.

g) Each lead-in conductor shall enter the building through a non-combustible, non-absorptive, insulating bushing slanting upward toward the inside, or by means of an approved device designed to give equivalent protection.

h) Each lead-in conductor shall be provided with an approved protective device (lightning arrestor) which will operate at a voltage of 500 volts or less, properly connected and located either inside the building at some point between the entrance and the set which is convenient to a ground, or outside the building as near as practicable to the point of entrance. The protector shall not be placed in the immediate vicinity of easily ignitable stuff, or where exposed to inflammable gases or dust or flyings of combustible materials.

i) If an antenna grounding switch is employed, it shall in its closed position form a shunt around the protective device. Such a switch shall not be used as a substitute for the protective device.

(It is recommended that the antenna grounding switch be employed, and that in addition a switch rated at not less than 30 amperes, 250 volts, be located between the lead-in conductor and the receiver set.)

j) If fuses are used, they shall not be placed in the circuit from the antenna through the protective device to ground.

k) The protective grounding conductor may be bare, and shall be of copper, bronze, or approved copper-clad steel. The protective grounding conductor shall be not smaller, nor have less conductance per unit of length, than the lead-in conductor, and in no case shall be smaller than No. 14, if copper, nor smaller than No. 17, if of bronze or copper-clad steel. The protective grounding conductor shall be run in as straight a line as possible from the protective device to a good permanent ground. Preference shall be given to water piping. Other permissible grounds are grounded steel frames of buildings or other grounded metal work in the building, and artificial grounds such as driven pipes, rods, plates, cones, etc. Gas piping shall not be used for the ground.

l) The protective grounding conductor shall be guarded where exposed to mechanical injury. An approved ground clamp shall be used where the protective grounding conductor is connected to pipes or piping.

m) The protective grounding conductor may be run either inside or outside the building. The protective grounding conductor and ground, installed as prescribed in the preceding paragraphs **k)** and **l)**, may be used as the operating ground.

(It is recommended that in this case the operating grounding conductor be connected to the ground terminal of the protective device.

If desired, a separate operating grounding connection and ground may be used, this operating grounding conductor being either bare or provided with an insulated covering.)

n) Wires inside buildings shall be securely fastened in a workmanlike manner and shall not come nearer than 2 inches to any electric light or power wire not in conduit, unless separated therefrom by some continuous and firmly fixed non-conductor, such as porcelain tubes or approved flexible tubing, making a permanent separation. This non-conductor shall be in addition to any regular insulating covering on the wire.

o) Storage-battery leads shall consist of conductors having approved rubber insulation. The circuits from storage batteries shall be properly protected by fuses or circuit breakers rated at not more than 15 amperes and located preferably at or near the battery.

Testing the Aerial and Ground Circuit

As it is rather inconvenient in many cases to inspect the aerial thoroughly, the owner who has trouble, such as crackling, howling, or whistling, can make a few simple tests, which will tell whether the trouble lies in the aerial and ground circuit, is due to atmospheric conditions, or lies in the receiver. One method of doing this is to disconnect the lead-in and ground wires from the receiver while it is in operation, and if the trouble ceases, it is presumably in the aerial or ground or is caused externally. If the trouble continues, it is most likely in the receiver or the accessories. However, there may be trouble in both the aerial circuit and the receiver.

A more accurate test of the aerial and ground is accomplished by means of a headset and a battery. A B-battery is used for this purpose, it being connected in series with the headset. Disconnect the aerial and ground wires from the binding posts on the receiver and place them in such a position that the bared ends of the wires will not touch each other, or touch any conductor such as metal which is connected to the ground.

To test for a grounded condition of the aerial, hold one tip of the tester to the aerial and touch the other to the ground wire or to any metal connected with the ground, such as a radiator or faucet, as shown in Fig. 40. If a click is heard, the aerial is grounded. A high-voltage battery of 45 or 90 volts should be used for making this test. Another method of testing for a grounded condition in the aerial circuit is to tune in a signal carefully and then disconnect the ground wire from the receiver. If the signal comes in louder after this is done, the aerial is grounded.

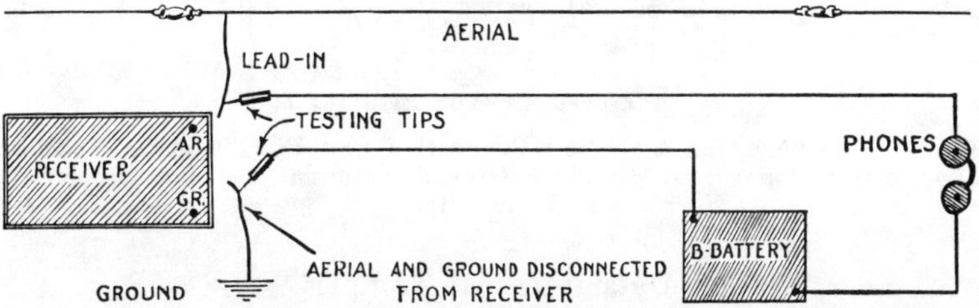

Fig. 40. It Is an Easy Matter to Determine Whether an Aerial or Lead-in Wire Is Grounded by Touching the Testing Tips to the Disconnected Lead-in and Ground Wires

In order to find out whether there is a break or an open circuit in an aerial, connect the far end to the ground, as shown in Fig. 41, and then make the same test as just explained. A click in the phones should be heard in this case, and the absence of a click indicates a break or open circuit. If this test is made on a windy day, when the aerial sways about considerably, and if crackling noises are then heard, one may be certain that there is a loose connection, probably at the point where the lead-in wire is connected to the aerial.

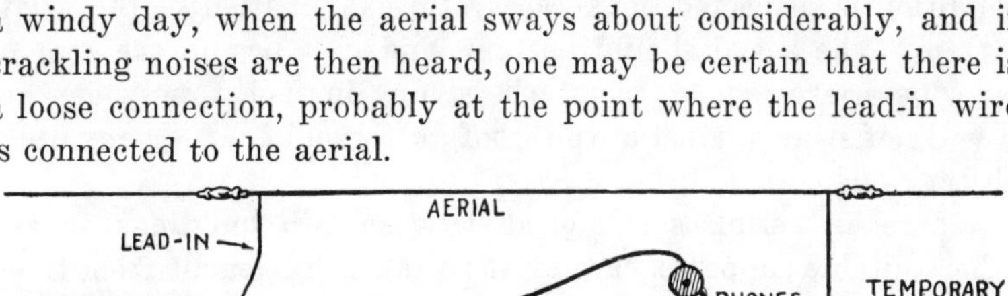

Fig. 41. By Connecting a Temporary Ground Wire to the Far End of the Aerial, and then Touching the Testing Tips to the Disconnected Lead-in Wire, One Can Determine Whether the Aerial Is Open-Circuited

The efficiency of a ground connection can be tested as shown in Fig. 42. A distinct click heard in the headphones shows that the ground is in good condition. Another method of testing the efficiency of a ground connection is to tune in the whistle of a broadcasting station, and then to touch, with a moistened finger, the bare ground lead where it connects to the receiver. If the whistle varies in pitch when this is done, the ground connection is inefficient.

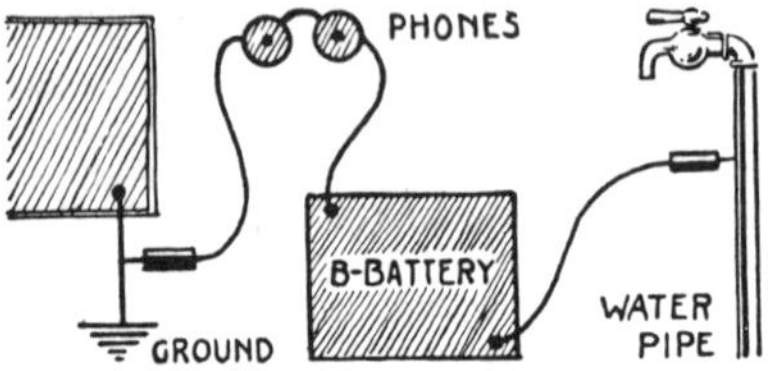

Fig. 42. Efficiency of a Ground Connection Can Be Tested by Touching One Testing Tip to the Receiver Ground Terminal and the Other to a Water Pipe

Outside Aerial Troubles

Grounded Aerials.—If the aerial touches any object or conductor that is connected or partly connected with the ground, the aerial will be grounded and poor reception will result, evidenced by a loss of volume or total inaudibility. Radio currents, which pass through the medium of the air, pass over wet walls, carbon-covered insulators very readily, and still more easily over metal such as conductor pipes. If an aerial is grounded or a grounded condition is suspected, make a careful examination of the whole system. The aerial should be taut, and if it lies on the roof, it must be re-erected. It should be strung in such a position that it will not sway against a wall, roof gutter, chimney, or any other object.

Where an aerial is stretched between two buildings or any other suitable supports, care must be taken to clear it from trees, which will ground the aerial, if they touch it. This is generally not taken into account when an aerial is erected during the winter months when trees are bare, with the result that the foliage often touches the aerial during summer. Trees may be used as supports for aerials, provided the end wire from the insulator to the tree is sufficiently long to permit the aerial proper to clear the branches and thus prevent a grounded condition.

Clean off the insulators thoroughly, as soot often collects on them. Soot is a form of carbon and permits the passage of electricity over the insulator, and some of the energy picked up by the aerial will then be lost. During winter, ice-covered insulators may also permit a loss of energy, but this trouble is usually periodic, and ceases as soon as the ice melts. In localities where there is considerable rainfall, loss of volume may be caused by a grounded condition, as a result of wet insulators. To avoid this trouble, a shield can be provided over each insulator, as shown in Fig. 43. This will act as a tiny umbrella, keeping the insulator dry. It is often stated that a receiver works better during a rainstorm than at any other time, but this assertion is erroneous. Bare lead-in wires should never be used, for their use greatly increases the possibility of grounding troubles. If a ground is found at a point where the lead-in wire touches a part of the building, and the insulation is frayed, provide a suitable support at this point, with an insulator on it to hold the aerial.

Sometimes faulty lightning arrestors will ground aerials. A lightning arrestor is really a spark gap having its two terminals

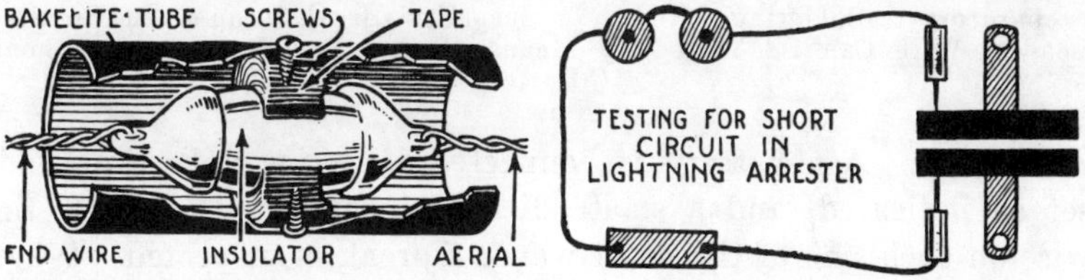

Fig. 43. Left: Bakelite Tube on Insulator Shields It from Rain and Prevents Radio Current from Being Grounded. Right: Lightning Arrestors Develop Short Circuits and Should Be Tested Occasionally in the Above Manner

set closely together. If the assembly becomes loose, the terminals may touch each other. The condition of the lightning arrestor can readily be checked with the headphones and a small C-battery, as shown in the right-hand detail of Fig. 43. Disconnect the leads from the lightning arrestor, and then proceed to test it. If it is in good condition, no click will be heard. On the other hand, a click indicates a short circuit. Besides short circuits inside of a lightning arrestor, there may be external losses through a film of soot. This condition is not really a short circuit, but nevertheless decreases the efficiency of the receiver. It is an easy matter to clean off the lightning arrestor with gasoline occasionally to prevent this trouble.

Open-Circuited and Broken Aerials.—Total inaudibility can sometimes be traced to an open-circuited aerial. The lead-in wire may be broken, or it may not have been soldered to the aerial. Corrosion at the joint may be responsible for an open-circuited condition. Look for an open circuit where the lead-in wire has been bent frequently, or where it has been mashed, as, for instance, under a window sill. A broken wire can usually be felt, but an accurate test can be made with the headset, as shown

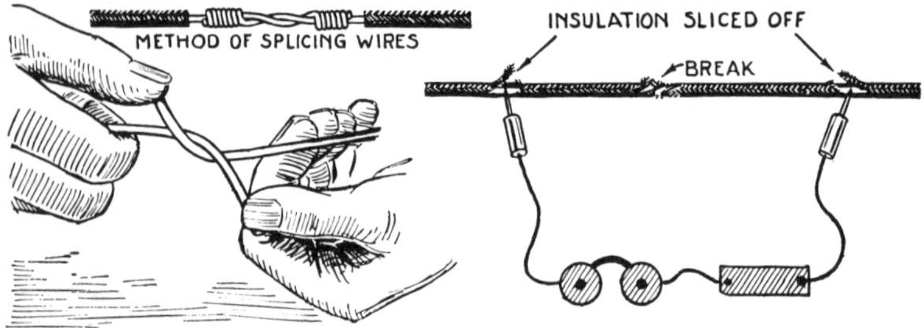

Fig. 44. Left: Proper Method of Splicing the Ends of Two Wires Together Preparatory to Soldering. Right: A Suspected Break in an Open-Circuited Lead-in Wire Can Be Tested by Means of the Headphones and a Small C-Battery

in Fig. 44. A C-battery is connected in series with the headset as indicated, and a small slice of insulation is cut off the wire on each side of the point where a break is suspected. When the testing tips are touched to the bare wire at the places where the insulation has been sliced off, there should be a sharp click, which indicates continuity of the wire. The absence of a click indicates an open circuit. Handle the wire at the place where the break is suspected, for the two ends may happen to touch each other while making the test, and this sometimes gives the same effect as an unbroken conductor. In case the wire is broken, cut the insulation at the break and bare the ends. Scrape them clean to the copper and then twist them together to form a splice, taking care not to twist the separate turns of wire too close together. The splice thus made must be soldered and wrapped with tape. Any film of corrosion between the lead-in wire and the aerial must be removed, which is done by untwisting the lead-in wire and scraping both wires clean. Always solder such connections. A poor connection between the lead-in wire and the aerial causes crackling noises in the loudspeaker, especially on windy days, when the connection is agitated.

A broken aerial can readily be seen, and it is repaired by splicing the two ends together. When doing this it will be found convenient to untwist the aerial from one of the insulators. After the splice has been made the aerial is re-erected. When aerials are attached to trees, allowance must be made for a certain amount of play, for, as the wind blows, the tree bends, and an aerial stretched tightly from a tree may be pulled taut and be broken by the strain. The remedy is to provide a coil spring, such as an ordinary screen-door spring, between the insulator and the tree, which will keep the aerial taut and at the same time allow the necessary play. Another method is

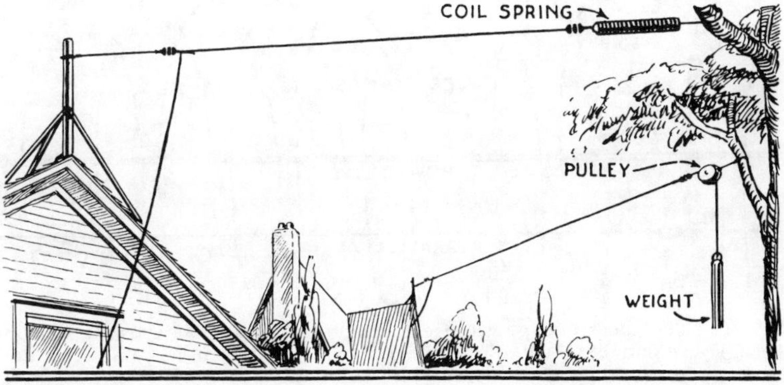

Fig. 45. Two Methods of Providing Play in Aerials Attached to Trees in Order to Prevent Undue Strain and Possible Breakdown, Caused by the Trees Swaying in the Wind

to suspend a pulley and provide a weight. Both methods are shown in Fig. 45. If an aerial sways considerably, the end wires that hold it may become brittle and break at the point where they are attached to the supports. To prevent an aerial from swaying, support it at one or more points. Weight on the aerial, such as is caused by ice during sleet storms, may cause it to break, but this trouble is rare, except in case of ribbon aerials.

Aerial Parallel to Power Lines.—When considerable trouble is experienced from a hum of regular pitch and frequency, and no alternating-current equipment is used with the receiver, the trouble may be caused by the aerial being parallel to a power line. Alternating-current surges in power lines set up a magnetic field at right angles to the conductor through which this current is flowing. This magnetic field consists of a great number of lines of force, radiating around the power line for a considerable distance. When they intercept other conductors parallel to the power line, they induce a slight current to flow in these conductors, the fluctuations of this current being in synchronism with the fluctuations of current in the power line. If such conductors happen to be radio aerials, it is evident that undesired currents are picked up by the aerials, and sent through the tuning inductance of the receivers connected to them. Such currents may be slight, but they are strong compared to radio-frequency currents, and when amplified by the tubes of the

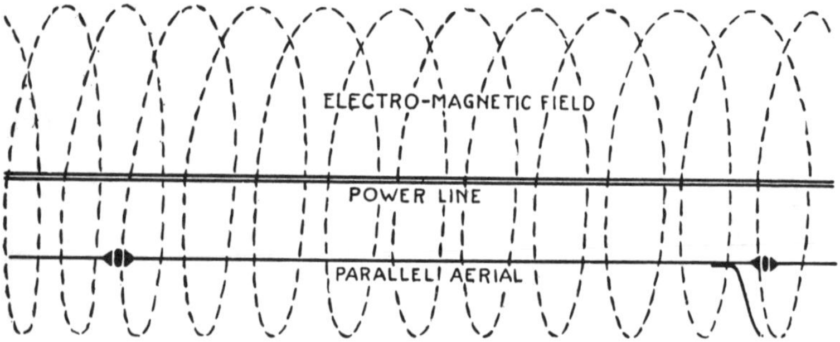

Fig. 46. The Electro-magnetic Field Produced by Electric Current Flowing through a Conductor. Radio Aerials Parallel to Such a Conductor Pick up Interfering Signals

receiver, they are very noticeable as a constant hum. Fig. 46 shows how an aerial intercepts the electro-magnetic field of a power line. Similar interference may also be experienced when an aerial is parallel to a telephone line, and a loud buzzing sound may sometimes be heard, caused by the bell-ringing apparatus used in telephone lines. When aerials are parallel to street-car lines, power-line hum may also be heard.

It is advisable to avoid erecting an aerial parallel to other aerials, for if any of such parallel aerials are used with regenerative receivers, considerable trouble may be experienced from reradiation. Interference from parallel power lines, telephone lines, street-car lines, etc., is less noticeable when the aerial is quite a distance away from them, but in some locations proximity to these lines is unavoidable. The best remedy for this trouble is to erect an aerial at right angles to the line causing the trouble. Then the lines of force will not intercept the aerial, and consequently no currents will be set up in it.

Indoor Aerials

As radio waves penetrate walls of buildings, there is practically just as much pick-up on an aerial, which is erected in the attic of a frame building, as there is on an aerial erected outdoors. But the objectionable feature of such an installation is that the aerial cannot usually be strung in a straight line. The pick-up value of inside aerials located in buildings having considerable steel framework is, however, much less than that of outdoor aerials, owing to the shielding effect and absorption of the steel framework. Inside aerials are particularly adapted for use with radio-frequency and neutrodyne receivers.

Attic Installations.—The advantage of an inside aerial is that no lightning arrestor is necessary, and the aerial is protected

Fig. 47. In Frame Buildings an Attic Aerial Installation as Shown Here Is Almost as Efficient as a Regular Aerial Erected above the Roof

from soot and corrosion caused by smoke and weather conditions. When making an attic installation, the shape and size of the attic must be taken into consideration. The end-to-end type of attic aerial is shown in Fig. 47; it consists of a stranded wire, looped through the eyes of insulators which are attached to the walls at both ends of the attic. The wire is pulled up tight and fastened to the end insulators, and the lead-in wire is soldered on. The total length of the aerial and the lead-in wire must be that which is most suitable for the receiver. The lead-in wire should be No. 14 stranded, insulated wire. It can be brought down through the partition to the location of the receiver, and a neat wall installation can be made at the receiver end, as shown in Fig. 36. Split, porcelain knobs used

Fig. 48. Indoor Aerial Can Often Be Used on Neutrodyne and Radio-Frequency Receivers. Left: A "Mat" Type of Aerial, Which Can Be Laid under a Carpet. Right: "Curtain" Type of Aerial Resembling a Tapestry

(Courtesy of "Popular Mechanics Magazine")

for electric-wiring purposes can also be used for holding the aerial in place, and they can be obtained at any electrical store. Where attic space is limited, the latter arrangement is recommended.

"Mat" and "Curtain" Aerials.—The hidden "mat" type of inside aerial may be laid on the floor and under the rug, as shown in Fig. 48. It consists of two sheets of heavy insulating paper, with the wire arranged between them, the paper sheets being fastened together securely with insulating glue. One end of the wire in the mat thus formed is connected to a binding post to which a lead from the receiver is connected. It will be found that such a mat will make a radio-frequency receiver much sharper than it is ordinarily when operated on an outdoor aerial, especially in cities where there are a number of broadcasting stations operating simultaneously. Another advantage of an aerial of this type is that there will be less trouble caused by reradiation if the receiver happens to be one of the regenerative type. The fact that this type of aerial can be hidden will be of interest to those who wish to keep as much apparatus out of sight as possible. No lightning arrestor is needed with this aerial.

The advantages of the mat type of inside aerial are also shared with the "curtain" type. Some of these come in the form of window shades, imitation tapestry curtains, and large pictures. A "tapestry" aerial is also shown in Fig. 48. It should be remembered, however, that in locations which are quite a distance from broadcasting stations, an outdoor aerial is usually preferable to an inside aerial unless the receiver is specially designed for an inside aerial, or works on a loop.

"Moulding" Aerials.—If the radio owner desires, he can install an inside aerial on the moulding of the room in which the receiver is located. Fig. 49 shows how this is done. Get some stranded, insulated wire, about No. 22 gauge, and start to lay the wire behind the moulding just above the receiver. It is usually necessary to use small staples to hold the wire in place, taking care, however, not to fray the insulation of the wire while fastening it in place, which may allow the bared wire to touch the plaster, and this may result in an appreciable loss, especially if the plaster is damp. The wire should be of a length suitable to the type of receiver used. It may be found necessary to bring the wire around the moulding more than once, depending on the size of the room. One end of the aerial is brought down to the aerial post of the receiver. A usual ground connection is required, but no lightning arrestor is necessary. For temporary use, an inside aerial may be arranged by simply laying the aerial wire on the floor underneath the carpet, in the form of a large coil, and bringing one end to the receiver.

Troubles Experienced with Inside Aerials.—Radio-frequency and neutrodyne receivers will usually work better on inside aerials than most other types of receivers. Small receivers, having two or three tubes, do not, as a rule, operate well on inside aerials, except in the immediate vicinity of a broadcasting station. Reduced signal strength and the inability to tune in distant

Fig. 49. Many Radio-Frequency and Neutrodyne Receivers Will Operate Satisfactorily on an Aerial Laid in the Moulding of the Room in Which the Receiver Is Used

stations is a trouble often caused by using an inside aerial on a receiver that does not have enough amplification. Another trouble frequently experienced with an inside aerial is that the receiver breaks into oscillation easily when the operator tries to obtain maximum volume. This trouble may also be caused by outdoor aerials of insufficient length.

Grounded conditions of inside aerials can also prevail; this is usually caused by a careless installation and the neglect to use insulators. It is not advisable to use bell wire for an inside aerial, as its insulation is usually thin and is easily frayed, exposing the bare wire, which may touch damp plaster and cause considerable loss or a completely grounded condition. Some radio owners merely drive nails into the plaster at various points in the room and hook the wire from one nail to another, twisting it around each nail to hold it in place. Such practice often bares the wire, allows it to make contact with the nail, and permits losses through the plaster. Never use a screwdriver or any sharp-edged tool to force the aerial wire behind the moulding for you may bare the wire and push it against the plaster, or perhaps break the wire, which would result in an open-circuited condition. The use of solid wire is also objectionable for this purpose, as it is bent so often while being handled, that it becomes brittle and is then likely to break.

Using Lighting Circuit for an Aerial.—A form of inside aerial, which has found much favor among radio owners, is a small screw-plug device, which, when attached to any light socket, utilizes the wiring of the house as an aerial system. Such a

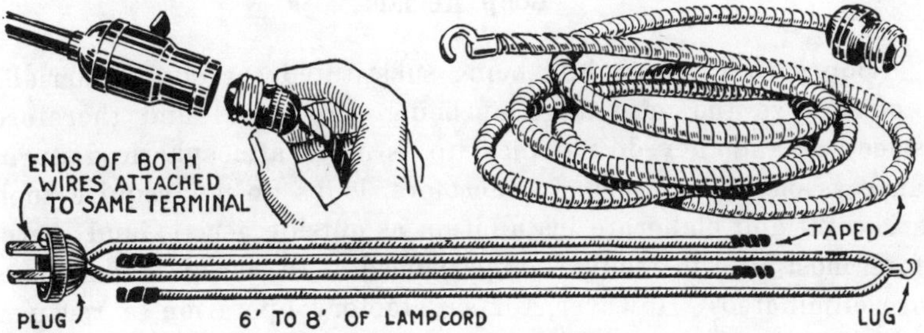

Fig. 50. The Electric-Lighting System of a Building Can Be Used as an Aerial by Providing an "Antenna Plug" or a Homemade Condenser, Which Consists of 6 to 8 Ft. of Lampcord, and a Plug for Connection to a Light Socket

device, shown in Fig. 50, is merely a condenser, one terminal of which is connected to one side of the line, and the other to the radio receiver, there being no connection between the receiver and the lighting circuit, which, of course, obviates the possibility of electric shocks. When using one of these devices, it will be found that reversing the plug in the socket sometimes gives better reception, as one side of the light circuit is grounded, and the other side will work better as an aerial system.

If the owner desires, he can readily make a device of this kind himself. All that is necessary is 16 ft. of ordinary lampcord, and a screw plug, arranged as shown in Fig. 50. Two lengths of single lampcord, each 8 ft. long, are connected at one end to one terminal of the screw plug. Two other lengths of lamp-cord, also 8 ft. long, are soldered at one end to a common lug. The four cords are assembled as shown in the drawing, with the plug at one end and the lug at the other, the whole arrangement forming a condenser. Tape is wound around the cords to bind them together into one cable. After assembling, the plug is screwed into a lighting socket, and the lug is attached to the aerial post of the receiver. By simply turning the socket key, as if to light a lamp, the device is brought in electrical connection with the network of wiring throughout the whole building. Surprising results can be obtained from this sort of aerial arrangement, although the signals may not be quite as loud as they are when obtained on a long outdoor aerial.

Loop Aerials

Loop aerials are rapidly being substituted for outside aerials. Loops have the advantage of being directional and therefore selective, and they do not pick up as much atmospheric disturbance as outside aerials. Furthermore, loops do not require such a careful and elaborate installation as outside aerials, and therefore most of the troubles experienced with aerial installations are eliminated. However, for satisfactory operation, a receiver which operates on a loop, must have one or more stages of radio-frequency amplification to compensate for the small pick-up value of a loop, as compared to the pick-up value of an outside aerial. There are very few three and four-tube loop receivers, and these are usually not suitable for distant reception. The great majority of loop receivers have five or six tubes.

How a Loop Works.—On most receivers employing a loop, signals can be partly tuned in by pointing the loop toward the station from which the signal is being radiated, as indicated by the position AB in Fig. 51. When the loop is pointed toward the station, the waves strike the loop from one side entirely, which results in a small current being set up, and consequently signal strength is then at its maximum. The signal can usually be tuned out entirely by turning the loop at right angles to the position AB, as indicated by the dotted lines. The reason for this is that the radio-frequency waves strike both halves of the loop at the same time, setting up currents in opposite directions, which neutralize each other, resulting in the total absence of a signal. This effect is more pronounced the farther the receiver is located from a broadcasting station. When the receiver is located in the immediate vicinity of the station, the signal cannot always be tuned out entirely by manipulating the loop only.

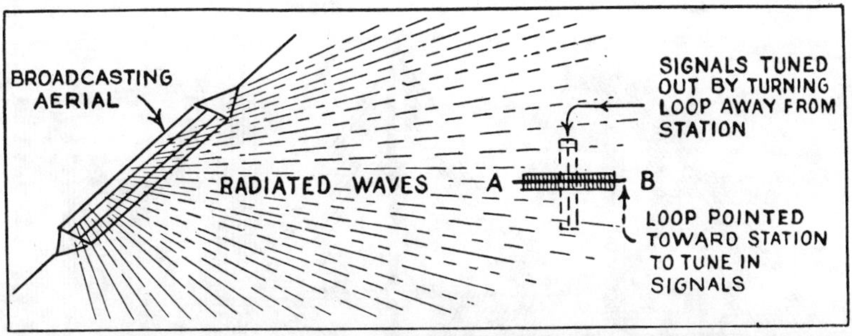

Fig. 51. Diagram Showing Directional Tendency of a Loop Which when Pointed toward a Broadcasting Station, as Indicated by *A B*, Brings in the Signals Loudest

"Box" and "Pancake" Types of Loops.—Various types of loops have been developed, but the types shown in Fig. 52 have proved to be the most popular. Detail 7 shows a "box" type of loop. It consists of a length of stranded, insulated wire wound on bakelite crosspieces attached to the arms. This method of winding seems to build up the signal strength to some extent, and these loops are therefore capable of picking up signals from greater distance and in greater volume than the "pancake" type, which is shown in detail 2. The winding on the "pancake" type of loop is merely attached directly to the arms as indicated. Selectivity is the chief feature claimed for this type.

Tapped Loops.—Tapped loops permit adjustment of their inductance by means of a small inductance switch. Tapped loops of this kind should not be confused with center-tapped loops; the latter have only one extra tap besides the two end terminals. Such a center tap is used on some receivers where a negative bias is required on the grid of the first tube, as is the case in some superheterodyne circuits having no grid condenser and leak on the first detector, which must then have a negative grid bias.

Loops of Adjustable Shape.—Another type of loop is the folding type, which can be changed from a narrow diamond shape to a square. This type of loop, however, has the disadvantage of making the logging of stations variable, as it has no definite points at which it can be set permanently, and when its shape is accidentally changed, the points on the dials where stations are tuned in are also changed.

Console Loops.—Obviously, large square loops are sometimes

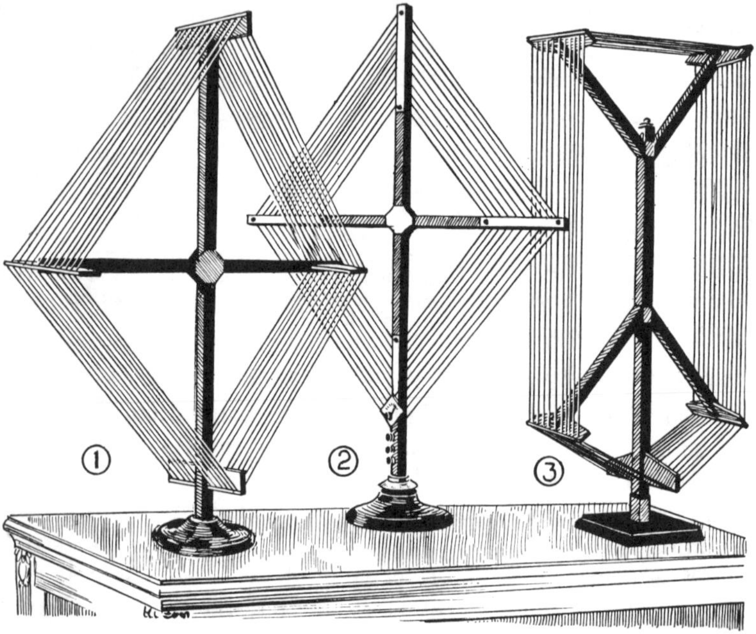

Fig. 52. (1) "Box" Type of Loop, Which Has Considerable Pick-up Value. (2) "Pancake" Type to Which Greater Selectivity Is Attributed. (3) Console Loop, Which Requires Little Space and Is Neat in Appearance

quite inconvenient in the home, especially where space is at a premium. To eliminate this objection, console loops of the type shown in detail 3 of Fig. 52 can be obtained. They are built in a rectangular shape, so that they can be placed on the receiver cabinet and can be rotated without touching the wall. Besides its convenience, such a loop excels the square loop in appearance.

Enclosed Loops.—To install the loop entirely out of sight is still better, but a smaller loop must then be used, and this means less signal strength, for the smaller a loop, the less will be its pick-up value, and vice versa. A method of installing a small loop in the lower compartment of a console is shown in Fig. 53. It must rest in bearings which will permit it to rotate freely, and some method of turning it by means of a dial or knob on the outside of the cabinet must be provided.

Fig. 53. Console Having a Small Loop Installed in the Lower Compartment. A Small Knob Connected to a Horizontal Shaft Is Used to Permit Rotation of the Loop in Any Direction

Selecting a Loop for Correct Inductance.—Loops are made in different sizes and shapes, and the selection of the right kind of loop for a receiver depends to some extent on the type of receiver. In general, it may be taken as a rule that the proper size loop to use, as far as its inductance is concerned, is one which gives good results over the whole waveband from 200 to 600 meters. If stations of 216-meter wavelength cannot be brought in on the tuning dial, there is too much wire on the loop, and if stations of 550-meter wavelength cannot be brought in, there is too little wire on the loop.

Loop Troubles

Much less trouble is experienced with loops than with outside aerials, chiefly because they are not subjected to the exposure of varying weather conditions. However, troubles do sometimes develop, and the radio owner can find them without much difficulty. The wire used on a loop and its insulation have considerable bearing on the efficiency of the loop. A good grade of stranded wire, covered with silk insulation, is usually considered the best, as silk does not have a tendency to absorb as much moisture from the air as cotton insulation. The smaller the size of wire used on a loop, the higher will be its resistance, and the pick-up value will sometimes be reduced to a noticeable extent. Bakelite supports for the wire are much better than wooden supports, as the moisture in the latter causes a considerable loss of energy.

Open-Circuited Loop.—As a result of constant rotation, the leads of a loop are frequently bent, which often results in a

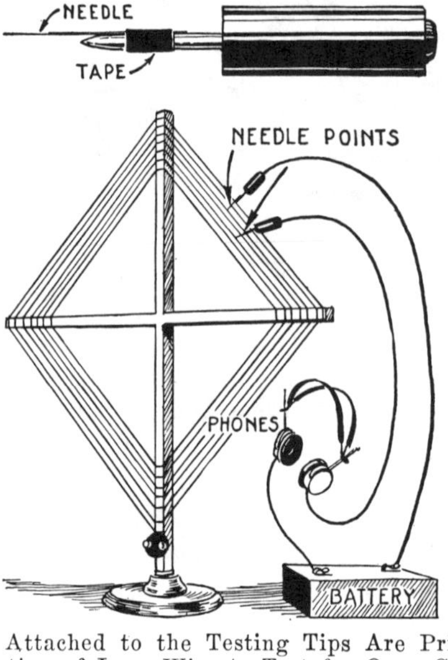

Fig. 54. Needles Attached to the Testing Tips Are Pricked through the Insulation of Loop Wire to Test for Open Circuit

broken connection. Loose connections will cause crackling noises, intermittent reception, or total inaudibility. When a break or open circuit is located, cut the wire at this point, twist the ends, and solder them together. After soldering, apply tape or collodion, if available. If the break cannot be located readily, fasten the tips of the headset tester to two darning needles with tape, and then proceed to make tests as shown in Fig. 54. The absence of a click denotes that the break is located between the points where the needles touch the wire. As the needles are sharp, they can be pushed through the insulation to make contact with the wire underneath without causing damage to the insulation.

Loose and Corroded Loop Connections.—On many receivers operating on loops, midget or cord-tip plugs and jacks are used instead of binding posts, as indicated in Fig. 55, and trouble is

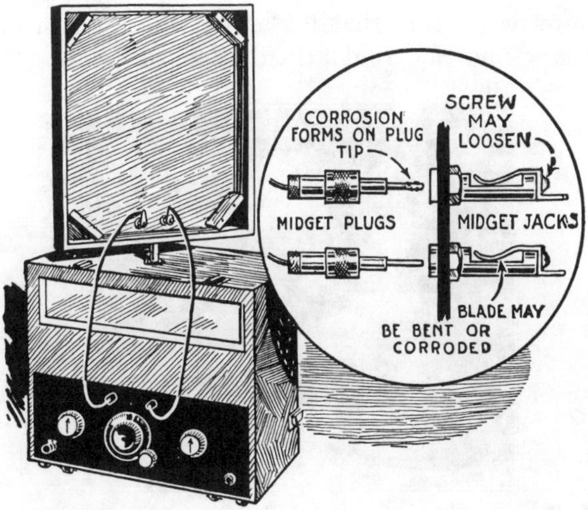

Fig. 55. Midget Plugs and Jacks Should Be Cleaned with Fine Sandpaper or Emery Cloth Occasionally. The Detail on the Right Shows Various Troubles to Which Midget Plugs and Jacks Are Subject

often experienced with them. The spring blade on the jack may become loose, or it may be bent slightly, so that a faulty electrical contact will result, which is evidenced by the same symptoms as open circuits, explained in the preceding paragraph. The remedy is to tighten up the screw that holds the blade in place, or bend the blade so that it will bear down on the cord tip securely, when the latter is inserted.

Sometimes the surface of the blade is covered with a film of corrosion, which is apt to form, especially if the receiver has been standing in a damp place for some time. The corrosion should be carefully cleaned off, which is done by removing the blade and rubbing it with a piece of fine sandpaper or emery cloth. It is a good idea to clean the cord tips in the same way. When trouble is experienced from corroded cord tips and jack blades, reception will be faint, and if the corrosive film is thick enough there will be no reception at all.

Loop jacks of the kind shown in Fig. 56 are used on some types of receivers. They are, however, apt to give the same troubles as the small cord-tip plugs and jacks. Examine the contacts, and sandpaper them occasionally in order to maintain the efficiency of the receiver. Lack of volume owing to these causes is similar to that caused by low batteries and paralyzed tubes, and if the owner is sure that his batteries are all in good condition, he may find the trouble in the loop contacts.

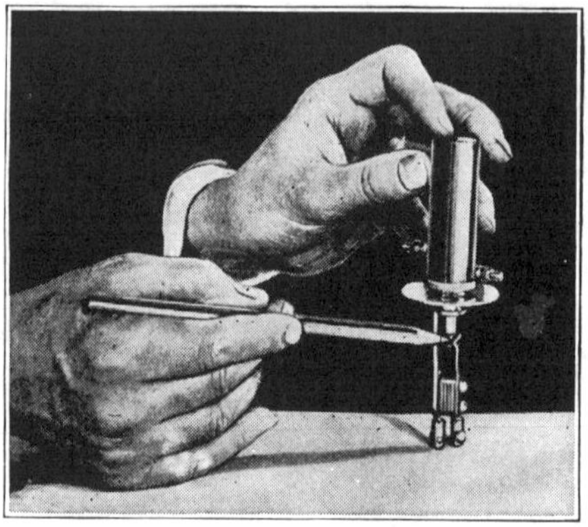

Fig. 56. Pencil Points to the Spring Blade on a Loop Jack, Which often Develops Trouble

(Courtesy of "Popular Mechanics Magazine")

Blanketing Effect of Steel Buildings.—Trouble is often experienced with loop receivers used in large office and apartment buildings, having considerable steel framework. Receivers, which work satisfactorily in other locations, may be entirely unable to pick up signals when they are installed in steel-constructed buildings. This is due to the absorptive characteristic of steel and iron for radio-frequency waves, which are electro-magnetic. When radio waves are intercepted by such buildings the waves are concentrated in the steel columns and grounded. Trouble of this kind is worse in some cases than it is in others, depending on the amount of steel in the building, and, of course, to some extent on the proximity of the broadcasting station. If the signals are not entirely blanketed, the steel columns nevertheless have considerable bearing on the directional quality of the loop; when tuning in a station, it will be found that maximum volume is obtained when the loop is pointed toward the center of gravity of the building. If the loop happens to be set near a steel column, this will sometimes affect its directional tendency. Proximity to a radiator, or to the lead-in wire of an outside aerial, may also cause a similar deflection of the loop. The only remedy that will effectively overcome this trouble is to erect an outside aerial. However, a loop receiver usually does not work on an outside aerial unless changes are made in the wiring.

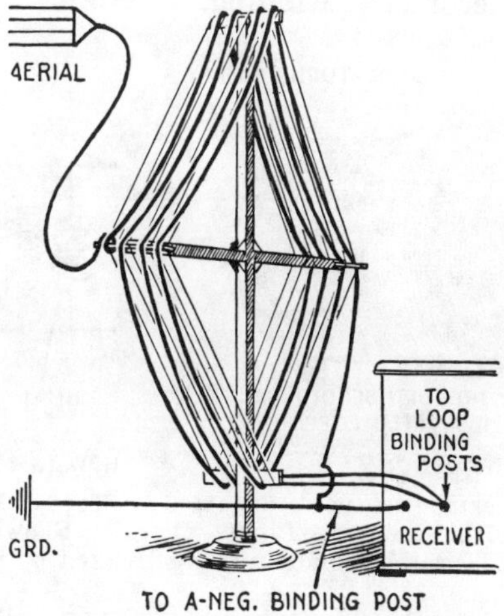

Fig. 57. A Loop Receiver Can Be Used with an Outside Aerial by Winding the Lead-in Wire Around the Loop

Changing a Loop Set to Work on an Outside Aerial.—Aerial receivers have an inductance coil tuned by means of a variable control, but loop receivers do not usually make use of such an inductance, as the loop itself serves as such. Loop receivers can, however, readily be adapted for use on outside aerials by making a few simple changes. The simplest method of doing this is to wind the lead-in wire around the loop a few turns, attaching the end to the ground and to the negative filament line of the receiver, as shown in Fig. 57. Another method consists in connecting to the receiver an inductance coil having a primary and secondary winding, as shown in Fig. 58.

Fixed inductance coils can be purchased at any radio store,

but one can readily be made. Get a cardboard tube 2½ in. in diameter. Coat the outside with shellac, and punch two small holes through the wall of the tube to hold the ends of the winding. Get ¼ lb. of No. 26 enameled or silk-covered copper wire, and proceed to wind this around the cylinder tightly until you have exactly 110 turns. Then wrap a layer of ordinary friction tape over the center of the secondary winding and wind 10 turns of wire over the tape. Apply another layer of tape to hold the primary in place securely and connect the four ends as indicated, providing a .0005 mfd. variable condenser across the terminals of the secondary winding.

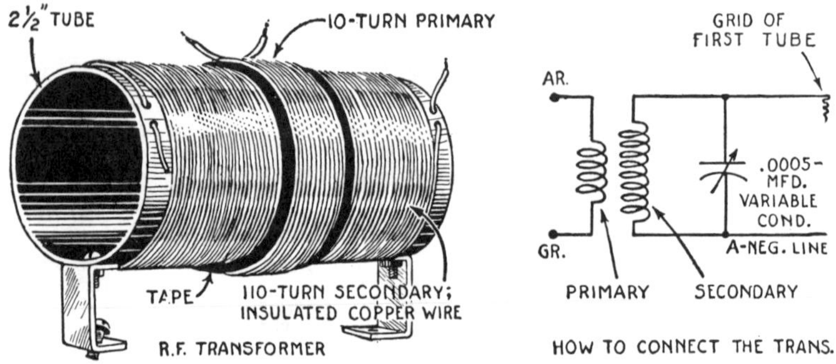

Fig. 58. Left: Homemade Radio-Frequency Transformer Wound on a Bakelite Tube Provided with Brackets. Right: Diagram Showing How an R. F. Transformer and Variable Condenser Are Hooked up in the Circuit

After the change has been made, tune in a station having a wavelength of about 545 meters. This station should come in when the tuning dial is set at approximately 90. If a station having a lower wavelength than 545 meters seems to be above the range of the dial, then there is too much inductance in the coil. Take off a few turns of the secondary, so that the station will come in at about 90 on the dials. If the station comes in much below 90 the coil has too little inductance, and the secondary winding will have to be rewound so that there will be more turns. However, it has been found that 110 turns is about right on a 2½ in. diameter.

Inductive Effects from Loop.—Often whistles and uncontrollable oscillations are experienced in loop receivers. There may be various causes of this trouble, as for instance too high a plate voltage on the radio-frequency tubes; intercoupling between stages; external interference due to reradiation from regenerative receivers, etc., but it can also be caused by having the loop too close to the receiver, unless the receiver is completely shielded. The reason for this trouble is that the loop picks up some of the amplified radio-frequency currents and passes them to the grid of the first tube; it is really nothing more than too much regeneration, which cannot easily be controlled. The remedy for this trouble is to move the loop away

Fig. 59. This Webbing for the Loop Leads Prevents Variations of Dial Readings Due to Varying Mutual Inductance of the Leads
(Courtesy of "Popular Mechanics Magazine")

from the receiver. Shielding the inside of the receiver cabinet as shown in Fig. 220 will also be found effective. When the loop is moved farther away from the receiver, it will be necessary to use longer leads, and this will change the inductance value, with the result that the tuning-dial readings on the receiver will be changed. When substituting longer loop leads, arrange them so that they will be in a fixed position, for if they are moved about, the dial readings on the receiver may vary. Variations resulting from this cause are only slight, but slight variations in dial readings count considerably when tuning in for distant stations. Loop leads can be kept an even distance apart by means of two strips of heavy cloth, doubled and sewed together to form three ''conduits'' for the leads, as shown in Fig. 59.

Changing an Aerial-Designed Receiver for Use with a Loop

Radio owners who have receivers designed to operate on outdoor aerials can readily make changes in the wiring of the receiver to accommodate it to the use of a loop. However, it must be remembered that all aerial-designed receivers will not work satisfactorily on a loop, unless additional stages of radio-frequency amplification are provided ahead of the detector tube. Such changes require considerable rewiring, and in most cases there is no space for the additional instruments. It is not advisable to change an aerial-designed receiver for use with a loop, unless it has two or more stages of radio-frequency amplification. In the latter case the change can be made as shown in Fig. 60.

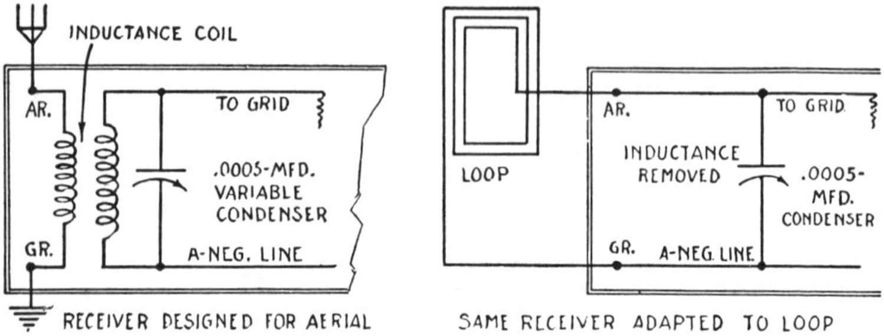

Fig. 60. A 5 or 6-Tube Receiver Designed to Operate on an Outside Aerial Can Usually Be Used with a Loop if a Few Simple Changes Are Made

Remove the tuning inductance coil from the receiver and substitute a .0005-mfd. variable condenser, connecting it across the lines that lead to the aerial and ground binding posts, which are then connected to the loop. One of these lines is run to the grid of the first tube, while the other is connected to the negative filament line. It will be found that a loop will give good reception with less interference than is experienced when using an outside aerial.

Adding a Radio-Frequency Unit to a Receiver

To compensate for the loss of volume resulting from changing an aerial-designed receiver for use with a loop, a stage of radio-frequency amplification in a separate unit can be added. It consists of a radio-frequency coil of the kind shown in Fig. 58, two .0005-mfd. variable condensers, a tube socket, and a rheostat. The latter must have a resistance of 6 ohms if a UX 201A tube is used, and a resistance of 25 ohms if a UX 199 tube is used. The unit is wired according to the diagram shown in Fig. 61, and all the connections must be well soldered. It is advisable to build this unit in a 6 or 9-in. radio cabinet, provided with a bakelite or hard-rubber panel. The unit is placed ahead of the receiver, and connection is made to the loop, receiver, and batteries as indicated. The tuning unit of the receiver, as indicated by the dotted lines, is removed, and connection is made directly from the aerial post of the receiver to the grid terminal on the first tube socket.

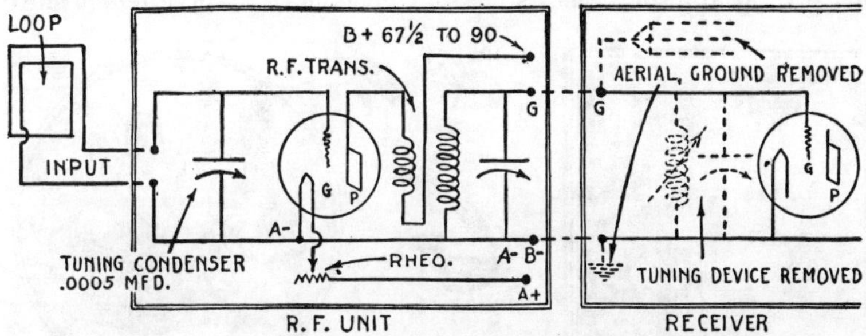

Fig. 61. When an Aerial-Designed Receiver Is to Be Used on a Loop, It Is often Necessary to Provide an Additional Stage of Radio-Frequency Amplification

CHAPTER IV

BATTERIES AND CHARGERS

Although A and B-eliminators are becoming more and more popular among radio owners in general, the great majority of fans are still using batteries as a source of filament current for their receivers. This prevailing use of A-batteries for supplying filament current for radio receivers is perhaps due to the fact that at the present time there are very few A-eliminators that give entirely satisfactory results. A-batteries may involve much more

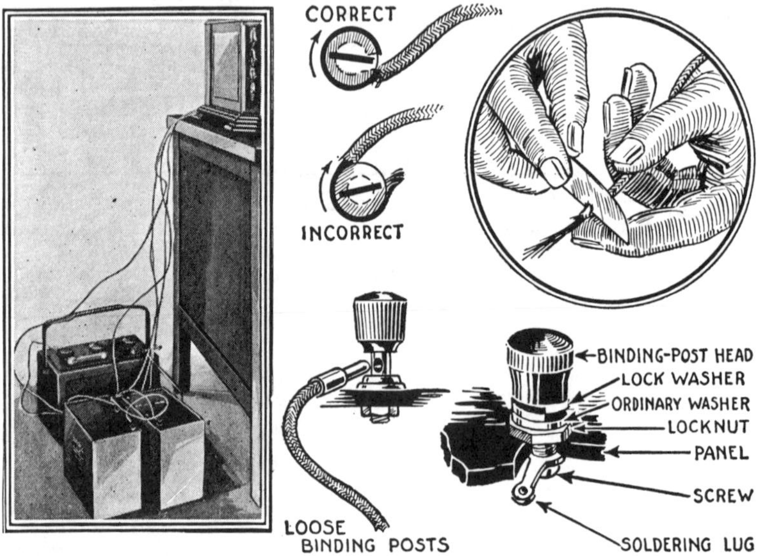

Fig. 62. Photo Shows a "Messy" Wiring Job. Details Show the Correct and Incorrect Methods of Attaching Wires under Screw Heads and also a Binding-Post Assembly

trouble and inconvenience than eliminators, because of the constant care necessary to keep them well charged, maintain the water level, etc., but on the whole, the results obtained with their use are superior to those obtained with many A-eliminators. The cost of eliminators is also a factor which restricts their general use. Besides, many radio owners do not feel justified in discarding a perfectly good battery and charger for new equipment, just to be up-to-date.

The large majority of radio troubles are caused by the condition of the batteries. This is not at all strange, as the entire operation of a receiver depends on its electrical supply; the A-battery providing the filament current, the B-battery providing the plate current, and the C-battery providing the grid-biasing voltage. The discharged condition of batteries is one of the most common troubles, but corroded terminals, acid-eaten wires, sulphated and warped plates, leakage of the electrolyte, corroding effects of electrolytic spray, run-down, noisy and internally open-circuited B-batteries, wrong connections, defective charger, etc., are some of the common troubles, which almost every radio owner may experience. A great many of these troubles can be prevented by simple means and proper care of the apparatus, which will be thoroughly explained in this section. When trouble is encountered in the receiver, and the owner is at a loss where to locate it, it is the best practice to examine the batteries and test them first, before attempting to examine the more difficult parts of the installation Besides, the batteries are more easily accessible than the other parts of the equipment.

Troubles in Battery Leads

Fig. 62 shows a very untidy battery installation, which detracts from the neat appearance of a radio set. Furthermore, an exposed installation of this kind is likely to develop trouble because the wires are in one's way when cleaning, and children can

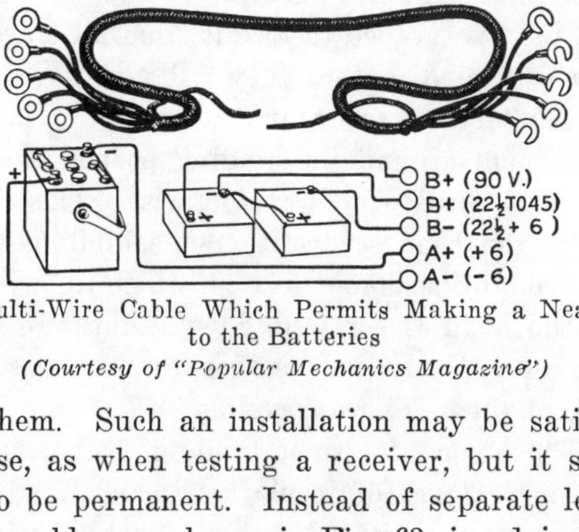

Fig. 63. A Multi-Wire Cable Which Permits Making a Neat Wiring Job to the Batteries

(Courtesy of "Popular Mechanics Magazine")

also get at them. Such an installation may be satisfactory for temporary use, as when testing a receiver, but it should never be allowed to be permanent. Instead of separate leads the use of multi-wire cables, as shown in Fig. 63, is advisable, and the batteries and charger should be kept in a closed box, in case the owner has no console.

Loose Binding Post.—When a receiver fails to operate and there is total inaudibility of reception, or considerable interference such as crackling noises is evidenced, examine the battery leads attached to the binding posts of the receiver. It is possible that the binding-post heads have become loose, perhaps from vibration, or they may not have been turned down tightly when the receiver was first installed. If the bared ends of the leads are not held down securely by the binding-post heads, the resulting loose connection may cause crackling, or in case the lead has slipped away from the binding post entirely, there will be no reception at all. If a filament lead happens to be loose at this point the tubes will flicker or fail to light. There is also the danger of a loose wire shifting over to the bared end of an adjacent wire or binding post, causing a short circuit, which may burn out the tubes, damage other parts of the receiver, short-circuit the battery, or cause it to run down rapidly, resulting in hot wires and possibly fire. The owner should go over the connections regularly to see that they are all tight. Stranded, insulated wire is the best to use for making connections, as it is much easier to handle than solid wire.

Single Conductors and Multi-Wire Cables.—The correct and incorrect methods of attaching the bared ends of the leads to the binding posts is also shown in Fig. 62. After the insulation has been removed from the end of the wire for about an inch, and the wire is scraped to free it from all traces of rubber, the strands are twisted so that they will not unravel or spread apart when the binding posts are tightened. The end of the wire should be bent around the binding post in the same direction that the head turns, which is clockwise. This tends to keep the wire under the head securely. Some binding posts have a hole drilled through the shank through which the wire is inserted, after which the head is screwed down tight. It is really the best practice to apply a little solder to the ends of stranded wires to prevent them from unraveling.

Do not use bell wire, for some kinds of bell wire are covered with only a single layer of insulation, which readily strips off and permits short circuits. Multi-wire cables with plug and socket attachments to facilitate connecting and disconnecting the receiver from the batteries quickly, as shown in Fig. 64, can be obtained for the purpose, and in many receivers they are already attached, the ends of the leads being connected to the wiring of the receiver permanently. The card of instructions for connecting the cable should always be kept handy, for it may be found useful at some time, especially if the leads are not marked with small tabs. It has become universal practice to use wires having red-colored insulation for carrying positive current. If the owner has been annoyed by the binding-post heads coming loose repeatedly, it is a good idea to use a washer and then a lock washer, if the binding-post head is removable, as shown in Fig. 62, or else one may resort to soldering the leads to the binding posts permanently. This will practically eliminate the danger of loose connections between the leads and the receiver binding posts.

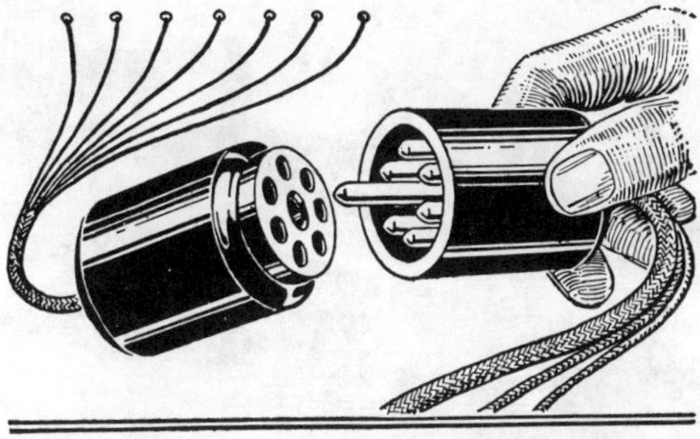

Fig. 64. Plug and Socket on a Multi-Wire Cable Permits Detaching the Leads from a Receiver Quickly

Useful Methods of Attaching Leads to Receiver.—Figs. 65 and 66 show two handy methods of attaching battery leads to receivers. In Fig. 65 a strip of bakelite or hard rubber is equipped with a number of midget plugs to fit into corresponding jacks which are provided on the receiver subpanel, the plugs and jacks, of course, being in perfect alignment. This method makes it possible to disconnect all the battery leads at once, and then there is no danger of the bare ends touching each other and short-circuiting the batteries. Of course, the same arrangement can also be used with leads from battery eliminators.

The second method, shown in Fig. 66, is also a safety measure against short-circuiting the leads, and is convenient as the leads can be disconnected from the receiver at any time. The bare ends of the leads are securely soldered to the inside ends of the tips on a discarded vacuum-tube base from which the glass shell has been removed. A socket is fastened to the receiver subpanel, and leads from the wiring of the receiver connected to it. Be sure that the receiver leads are all connected to the terminals on the socket in the correct relation to the battery leads attached to the tips on the tube base. By carefully following the illustration, which shows exactly where to attach the various leads, no mistake will be made in this respect. Use a

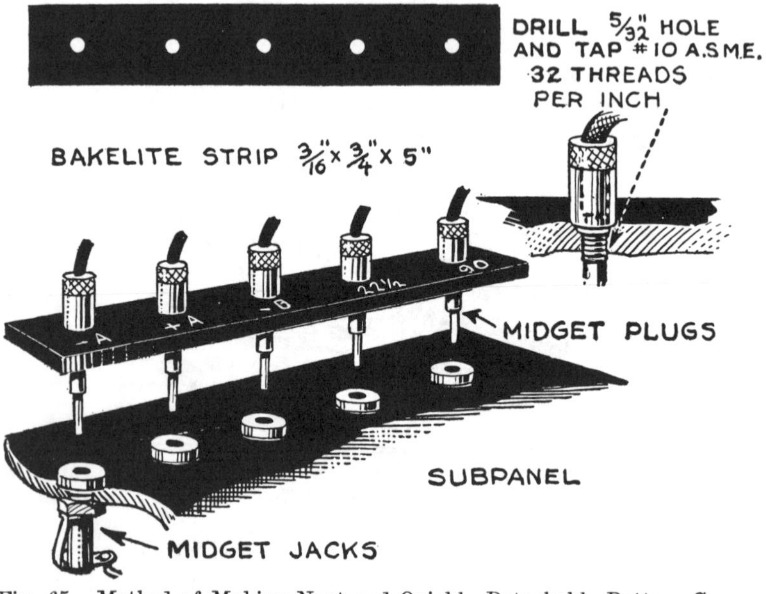

Fig. 65. Method of Making Neat and Quickly Detachable Battery Connections to a Receiver with the Use of Midget Jacks and Plugs

bakelite tube base instead of a metal base as the latter entails some danger of short circuits. In case of a small receiver where only two A-battery and two B-battery leads are used, one socket and base are sufficient. In case of a 5-wire installation, where the A-negative and the B-negative lines are connected, this connection can be made outside of the receiver, and then it is possible to run the A and B-negative current through one prong of the tube base, leaving two contacts for the B-positive connections and one for the A-positive. The same method can be used if the A-positive and B-negative lines are connected. In case of several B-positive taps, however, it may be necessary to provide two sockets instead of one.

Battery leads, which are loose at the ends where they are connected to the batteries, may cause the same troubles as mentioned above. It is a good practice to solder lugs to the ends of the leads that make contact with the batteries. As the lugs are flat, they will be held much more securely by the terminal heads

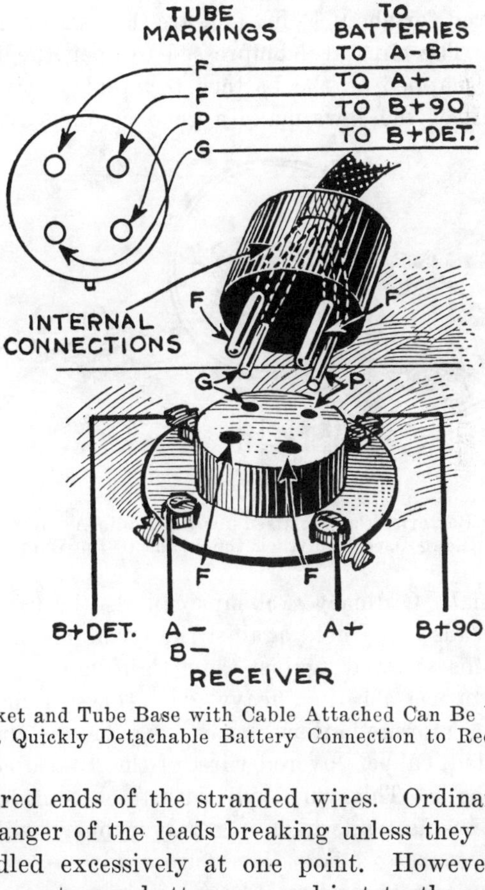

Fig. 66. Socket and Tube Base with Cable Attached Can Be Used for Making a Quickly Detachable Battery Connection to Receiver

than the bared ends of the stranded wires. Ordinarily there is not much danger of the leads breaking unless they are strained or are handled excessively at one point. However, the leads connected to a storage battery are subject to the ruinous effect of the battery acid, which eats through both insulation and wire, and gradually disintegrates it.

Storage Batteries

Attaching Leads to Storage-Battery Terminals.—On most storage batteries that are especially made for radio purposes binding posts are provided to facilitate attaching the wires. When using another type of storage battery, such as one intended for use in automobiles, one will have to improvise some method of attaching the wires to the terminals securely. Two methods of doing this are shown in Fig. 67. Spring clips for this purpose can be obtained at any store handling radio supplies. The ends of the wires are bared for about ½ in., and are then screwed or soldered to the clips. The clips are compressed to open the jaws so that they can be snapped on the battery terminals. Always be sure that the teeth of the jaws make a good contact with the metal

Fig. 67. Auto Batteries Can Be Used for the Radio A-Supply, Spring Clips or C-Clamps Being Used to Attach the Leads to the Battery Terminals

of the terminal. Ordinary C-clamps can also be used, the bared ends of the leads being held against the terminals of the battery, and the clamps screwed against them tightly.

It is best to use a rather heavy grade of stranded, insulated wire for the storage-battery leads, as, for instance, No. 14-gauge stranded, rubber-covered wire, of the kind used for house-wiring purposes. This wire will withstand considerable bending, and its insulation, which is partly rubber, will also stand up under exposure to acid. Never use a wire of small diameter for carrying filament current to the receiver, for the resistance of a wire increases as its size decreases, and increased resistance means less current-carrying capacity. Also, do not use excessively long A-battery leads, for the length of a wire increases its resistance.

Leads to Storage Battery Reversed.—As radio receivers are designed to operate only when the filament current flows through the instruments in a certain direction, it is essential that the filament leads must always be connected to certain terminals on the battery. If this order is reversed the receiver will operate unsatisfactorily, or perhaps not at all. To facilitate distinguishing between the battery terminals, one of them is usually marked POS., P., or +, and the other NEG., N., or —. In some cases the positive terminal has a red daub of paint to distinguish it from the negative terminal.

Determining Unmarked Battery Polarity.—As the terminals are made of lead, which is a very soft metal, the markings are easily effaced, and it often happens after a battery has been

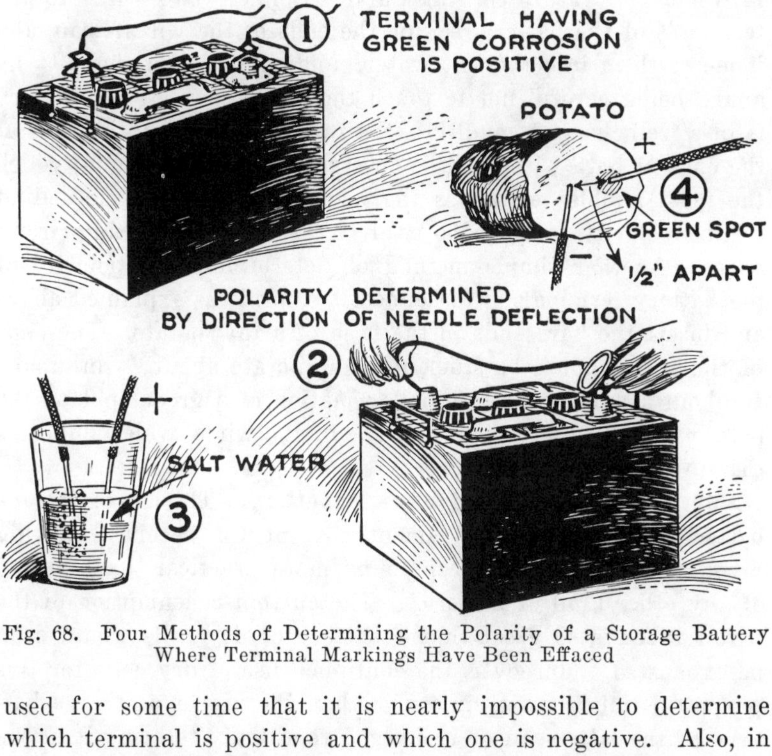

Fig. 68. Four Methods of Determining the Polarity of a Storage Battery Where Terminal Markings Have Been Effaced

used for some time that it is nearly impossible to determine which terminal is positive and which one is negative. Also, in case 6-volt auto batteries are used for radio purposes and the terminals are so corroded that their markings cannot be seen, it is necessary to use some method of determining the polarity. Various methods are shown in Fig. 68. If any traces of corrosion can be seen on the terminals, it will be found that the one having a green-colored corrosion is positive, and the one having a white or gray corrosion is negative. Another method of finding the polarity of a battery is to apply a suitable voltmeter to its terminals. The leads from the voltmeter are usually marked positive or negative, or else one lead may be red, indicating positive polarity. If the needle of the voltmeter deflects toward the highest reading, the leads of the instrument are touching battery terminals of the same polarity, while if the needle deflects away from zero and below the scale, the leads on the voltmeter are touching battery terminals of the opposite polarity. If no voltmeter is available, a simple test can be made in a glass of saturated salt water. Connect two wires to the terminals of the battery, scrape the ends of the wires clean, and immerse them into the salt water, holding the ends about ½ in. apart, being careful not to touch them together. If the battery is in a well-charged condition, the formation of small bubbles at the ends of both wires will be observed. The wire around which the most bubbles collect is the negative one. This method of testing polarity can also be used in case of d. c. lighting-circuit wires. Another simple method of determining the polarity of the battery terminals is to connect two leads as explained above and insert the bare ends in the flesh of a raw potato. The ends of the wires should be stuck into the potato about ½ in. apart. In about a minute the gradual formation of a green spot on the potato will be observed around one of the wires, which indicates that this wire is positive.

Internal Construction of Storage Battery.—The use of storage batteries for supplying filament current for receivers having 5-volt tubes has so far been found more practical than the use of any other kind of batteries. The current consumption of the 5-volt tubes is so great that a battery of dry cells would soon be exhausted. Moreover, the continued use of dry cells for this purpose is much more expensive than the use of a storage battery, as the latter can be recharged repeatedly to bring it in the same condition as that of a battery of dry cells when new.

A storage battery consists of three individual units or cells, each of which is capable of delivering approximately 2 volts of electrical pressure and a considerable current or amperage. The cells are connected in series, so that a total potential of 6 volts or slightly more can be obtained from the battery terminals. Each cell consists of a number of alternate brown and gray-colored plates, which are made of a lead-alloy framework or grid, holding a hardened paste. The brown plates are positive and the paste on these plates is peroxide of lead; the gray plates are negative, and the paste on these plates is sponge or pure lead. All the positive plates in each cell are connected together and similarly all the negative plates, making two sets of plates.

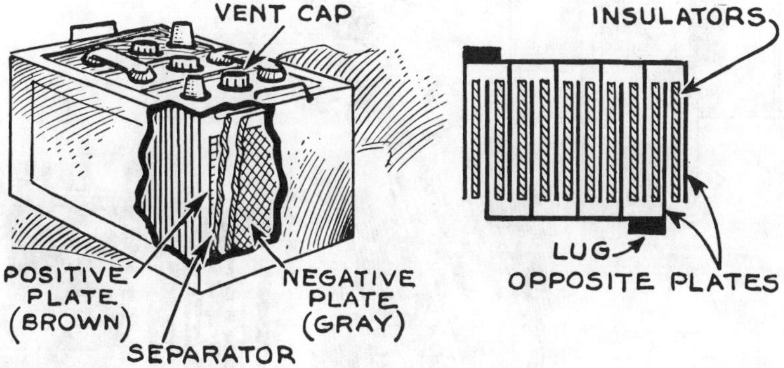

Fig. 69. Left: Internal Construction of a Storage Battery Showing Gray and Brown Plates with Separators Between. Right: Diagram Showing How the Plates Are Connected Together

as shown in Fig. 69. Each plate is separated from the adjacent one of opposite polarity by means of a wood or hard-rubber insulator. The whole plate assembly is set in a jar of glass or hard rubber, or other acid-resisting material, and the top of the jar is sealed shut with the exception of a small vent hole into which a cap screws. The terminal of each set of plates projects above the cell, and both are connected to terminals of opposite polarity on the adjacent cells. A solution of sulphuric acid and pure water is poured into each cell. A 5 to 1 solution of sulphuric acid and distilled water is customarily used as an electrolyte for storage batteries. The three cells of a 6-volt storage battery are set side by side in a heavier box which protects them against injury. Fig. 69 shows the internal construction of a storage battery.

How the Storage Battery Works.—Although there is a general belief that a storage battery literally stores up the electricity which passes through it during the charging process, this is not actually the case. When a battery is being charged, the electric

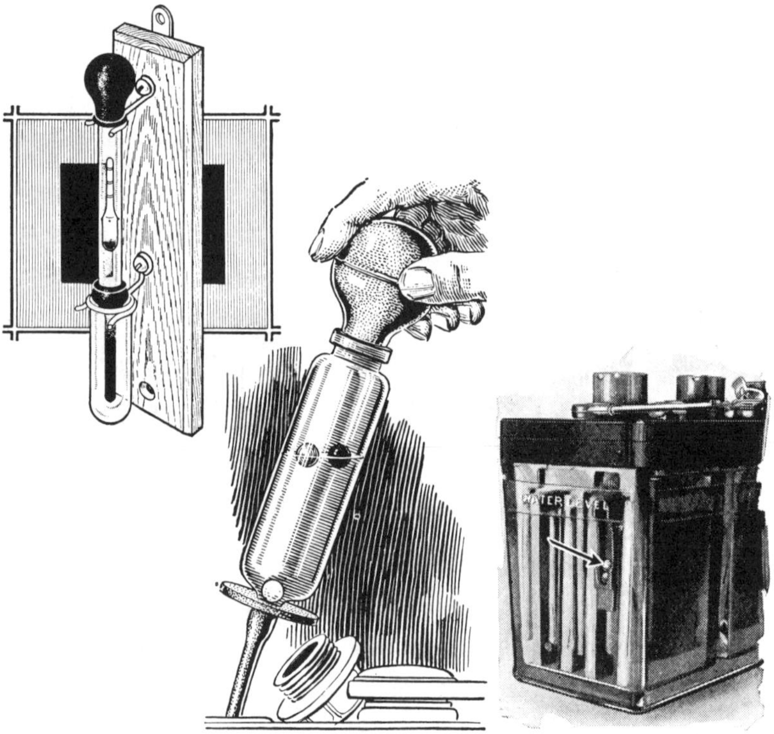

Fig. 70. Left: Usual Type of Hydrometer and an Easily Made Holder. Center: Hydrometer Having Balls of Various Density Values, Which Indicate the Condition of Charge. Right: Ball Hydrometer Incorporated in Battery Wall

(Courtesy of "Popular Mechanics Magazine")

current passing through it causes chemical changes to occur inside of it and these chemical changes are reversed when the battery delivers current. The density of the electrolyte of a storage battery changes along with the charge on the plates, and by taking tests of the electrolyte, the condition of charge can be accurately determined.

Testing Storage Battery with Hydrometer.—The best and most dependable method of testing the condition of charge of a storage battery is by means of a battery hydrometer. The most common kind is shown in Fig. 70, left detail. This instrument can be obtained at any auto-supply store, hardware store, or radio shop for 50 cents to $1.50, depending on its quality. It is a glass tube about 6 in. long, with a rubber bulb at one end and a rubber nozzle at the other end. In use, the nozzle is submerged in the battery electrolyte, the bulb is compressed and released to draw the electrolyte up into the glass tube. A long, graduated float, weighted at one end, is contained in the glass tube, and rises as the electrolyte is drawn up into the tube. The reading on the float closest to the surface of the electrolyte tells the exact condition of charge of the battery. The electrolyte will raise the float so that it reads from 1,275 to 1,300 when the battery is completely charged, old batteries showing a lower reading than new ones, but this is no cause for alarm. The reading will be about 1,215 when a battery is half charged, and about 1,120 when it is nearly discharged. Lower readings than this can sometimes be obtained, but the battery should never be allowed to drop so low. The hydrometer shown in Fig. 70, center detail, has two balls of different density. If both balls float on the electrolyte, when this is drawn up into the syringe, the battery being tested is fully charged. If one ball floats, the battery is half charged, and if both balls sink, the battery is discharged. Fig. 70, right detail, shows a storage battery having a glass jar in which such a hydrometer is incorporated

To take a hydrometer reading of a battery, first unscrew the vent caps of all the cells, and insert the nozzle end into the electrolyte of the first cell to be tested, pinching and releasing the bulb to draw the electrolyte up into the tube. After the reading has been taken, compress the bulb again so that all of the electrolyte runs back into the cell from which it was taken. Taking electrolyte out of one cell and returning it to another results in a weaker electrolyte in the first cell and a stronger electrolyte in the second, making subsequent hydrometer readings unreliable, owing to the variation in density of the electrolyte in the different cells.

If the level of the electrolyte is so low that the solution cannot be drawn up into the hydrometer, add some distilled water to bring the level about ½ in. above the plates. Do not take a hydrometer reading directly after water has been added, but charge the battery for several hours first, which will thoroughly mix the added water with the electrolyte and insure a correct reading. When the reading of the float is taken, care should be taken not to pinch the bulb, for this may cause an incorrect reading. All three cells of a storage battery should be tested, for the test of one cell does not necessarily indicate the condition of the others. No other device for testing storage batteries is as reliable as a hydrometer. Voltmeters and solenoid devices are not entirely dependable for this purpose.

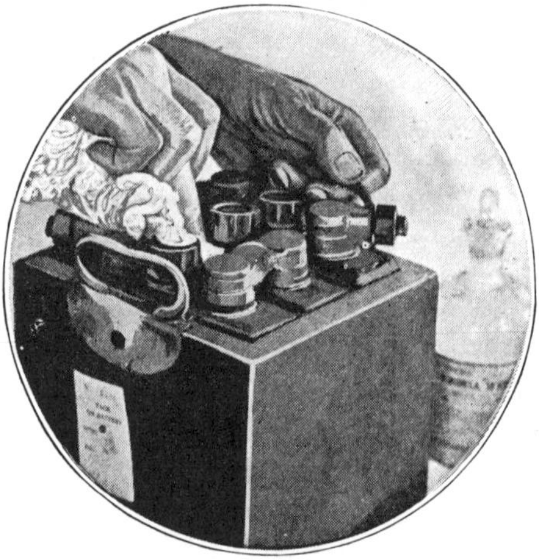

Fig. 71. Cleaning Electrolytic Spray from the Top of a Storage Battery with Ammonia-Saturated Rag

(Courtesy of "Popular Mechanics Magazine")

When transferring the hydrometer from one cell to another, be sure not to spill any of the electrolyte on the top of the battery for in such a case there will be a loss of acid and this may also provide a path for electrical leakage between terminals. Always keep the top of the battery clean and dry. This is done by first wiping it off with a rag moistened with ammonia or vinegar, and then with a rag saturated with plain water, after which a dry rag is used to remove the water, as shown in Fig. 71.

After all three cells have been tested and all the electrolyte has been squeezed out of the hydrometer, take a wide-mouthed bottle, such as a fruit jar or a quart-size milk bottle, containing some distilled water, hold it over the battery, and insert the hydrometer into it so that none of the electrolyte will drip on the floor, while making this transfer, as shown in Fig. 72. Such a bottle provides a good place in which to keep the hydrometer. If, while taking the readings of the cells, it is discovered that they vary a trifle, the battery should be given a slow overcharge at a rate of not more than ½ ampere, and this overcharge should be continued for 24 hours. This will usually equalize the reading of the cells, provided their electrolyte has the same density.

Ruinous Action of Electrolyte.—The hydrometer nozzle should not be raised out of the cell while the reading is being taken, as a drop or two of the electrolyte might get on one's clothes or on

Fig. 72. Drops of Electrolyte Ruin Rugs and Furniture, and It Is therefore Advisable to Take the Utmost Care When Using the Hydrometer

a rug. This will be disastrous, for this acid eats through cloth very readily. The same care should be observed when transferring the hydrometer from one cell to another. In case some of the electrolyte is spilled accidentally, a rag saturated with ammonia or vinegar should be quickly applied to the spot on which the electrolyte was spilled, or a liberal quantity of ordinary baking soda should be sprinkled on the affected spot. Ammonia, vinegar, and baking soda have a counteracting effect on the electrolyte; they tend to neutralize it and arrest its ruinous action.

Troubles Resulting from Low Level of Electrolyte.—Constant charging and discharging of the battery causes some of the water contained in the electrolyte to evaporate, and if the lost water is not replaced the level of the electrolyte will drop below the level of the plates, which may injure them. It is therefore necessary to keep the level of the electrolyte fully above the top of the plates, about ½ in. being advisable. Only distilled water should be added from time to time, as the sulphuric acid in the electrolyte does not evaporate to an appreciable extent. Usually water is added once a week during warm weather and every ten days or two weeks during cold weather.

Fig. 73. Set of Battery Plates Ruined by Neglecting the Level of the Electrolyte

One of the results of neglecting the water level so that it drops below the top of the plates is shown in Fig. 73. In this case the battery was neglected until about half of the electrolyte had evaporated, with the result that the plates were so badly sulphated that they had to be replaced. Such a sulphated condition permits only half of the plates to function, which means a great loss of efficiency. If it is found that the top of the plates of the battery has been exposed for a short time, there may be a slight sulphation; this can usually be remedied by giving the battery a slow overcharge, at a rate of ½ ampere or less for two or three days. A trickle charger will do the work nicely, but in case only a charger passing from 1 to 5 amperes is available, it will be necessary to insert a variable resistance, such as a radio rheostat and a small ammeter in series with it, as shown in Fig. 74. The rheostat can then be adjusted until the desired rate of charge is obtained. In case of a badly sulphated condition, it is best to take the battery to a reputable battery-service station for repair.

Harmful Effect of Impure Water in Storage Battery.—Water from the city mains, or rain water caught in metal pans, or water containing any minerals or salts should never to used for storage batteries, as such water contains matter which is injurious to the battery. City water conducted through metal pipes and water held in a metal container absorbs some particles of metal, and when this water gets into a battery, a slight coating of metal is deposited on the plates, which tends to decrease their efficiency. Water containing salt and minerals also has the same effect. Pure rain water may be used, provided it is caught in the open, in non-

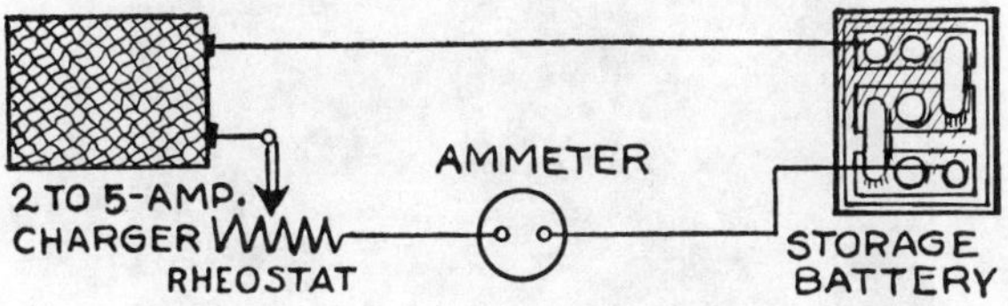

Fig. 74. Proper Method of Connecting an Ammeter and a Rheostat in Series with the Storage Battery to Control the Rate of Charge from a 2 to 5-Ampere Charger

metallic containers, and is not taken from roof gutters or conductor pipes.

Distilled water is the safest to use for this purpose, and it can be purchased at any drug store for a few cents per gallon. It must be kept in a glass or porcelain bottle or container, and should never be poured into a metal pan or dish when transferring some of it to the battery. It will be found convenient to pour some of the water into a porcelain cup or a glass tumbler, and to use a small rubber syringe for transferring the water to the battery. The hydrometer should never be used for a syringe to fill the battery with water, as the float may break from the jarring it receives under such usage. Too much water is, of course, also undesirable, for when the battery is being charged, the electrolyte gets warm and expands, which may cause it to run over and cause considerable damage.

Danger of Explosion.—When inspecting the battery to see whether or not it contains enough electrolyte, never hold a lighted match over the vent holes of the cells, as a serious explosion may result. When a battery is charging, hydrogen gas is liberated, especially at the end of the charge. This gas fills the space in the cells above the electrolyte. It is highly inflammable, and if ignited, may cause considerable damage. A flashlight, as shown in Fig 75, or an electric lamp on an extension cord should always be used rather than to take a chance with a match.

Fig. 75. To Inspect the Electrolyte Level with a Lighted Match Is Very Dangerous as an Explosion May Result. Use a Flashlight or an Electric Light for This Purpose

(Courtesy of "Popular Mechanics Magazine")

Keeping Storage Battery Well Charged.—For the most efficient operation of a receiver the storage battery should always be kept in a well-charged condition, and the charge should not be allowed to run down below 1,200. The first symptom of a run-down battery is a loss of selectivity on an otherwise sharp receiver. Considerable interfence may then be evidenced, and reception from distant stations will become an impossibility. As the battery approaches a discharged condition, the volume of reception will get weaker and the rheostats on the receiver will have to be advanced to restore the volume. But this helps only temporarily, and finally the tubes become dim, and no signals can be heard at all. To bring the battery back to normal considerable recharging is necessary, depending on the size of the battery. The charger should be kept going continuously until the hydrometer shows that the battery is again fully charged.

Precautions to Be Observed when Charging.—If a storage battery is kept in a console or on the floor under or near the

Fig. 76. When a Storage Battery Is Set on the Floor, a Rubber Mat Should Be Provided
(Courtesy of "Popular Mechanics Magazine")

receiver, it is a good plan to provide a glass tray or a rubber mat for it to stand on, as shown in Fig. 76, for during the charging, some of the electrolyte may bubble out. If it gets on the floor it will cause a dark stain, which cannot be removed, or it may ruin a rug. The electrolyte does not affect glass and rubber, and therefore glass or rubber trays are used. Before putting a battery on charge, be sure to unscrew the vent cap from each cell, for a battery produces gas while charging, and if the little pin hole in the vent cap is clogged with dirt, the gas inside sets up quite a pressure, which may finally force out the obstruction from the pin hole and permit the sudden escape of the gas, the suction also drawing out a spray of the electrolyte. To prevent this from occurring, with consequent ruin to rugs and furniture, the vent caps should always be unscrewed when charging.

When a battery on charge reaches the fully charged condition, it is likely to gas and bubble quite freely and to cause more or less spray, which settles on the top of the battery. After the charge is completed, the spray can be wiped off. A heavy charge can be put through a battery when it is low, but as it approaches the final stage of charging, when it is nearly full, a low, tapering rate of charge is preferable. Too rapid charging at this time may cause buckled or broken plates, as shown in Fig. 77. Rather

Fig. 77. When a Storage Battery Is Overheated from Charging It at an Excessive Rate, the Active Material Falls from the Plates and They May also Buckle

than to permit a battery to become completely discharged or nearly so before recharging it, it will be found that the most efficient practice is to keep the battery in a fully charged condition, by giving it a short charge every day, just enough to make up for the current that was used by the receiver. This method is better than charging for a long period once a week, and insures the most efficient operation of the receiver.

Corrosion on Battery Terminals.—Corrosion, which collects on the battery terminals, especially on the positive terminal, destroys within a short time the leads connected to it. This corrosion, copper sulphate, is either a green-colored paste or a chalky substance. A battery with badly corroded terminals is shown in Fig. 78. To prevent this corrosion from accumulating, the battery terminals should be carefully cleaned at regular intervals with a solution of ammonia and water, or a solution of ordinary baking soda and water in the proportion of one tablespoonful of soda to every 3 oz. of water. After the terminals have been wiped clean and dry, a coating of vaseline or ordinary cup grease should be applied. This coating will prevent subsequent corrosion. Do not apply the vaseline or grease to the terminals while the leads are disconnected, for after replacing them, it will be found that the film of grease is also an effective insulator, which prevents a good electrical contact of the wires to the battery terminals, with the

Fig. 78. Corroded Battery Terminals Can Be Prevented by Keeping Them Covered with a Coating of Vaseline

result that no current can be obtained. Corroded leads on storage batteries generally make poor contact, forming a high-resistance joint, which may cause crackling noises in the receiver or perhaps complete inaudibility.

General Maintenance of Storage Battery.—It is a good idea to take the storage battery to a battery-service station once a year to have it cleaned out and examined. By constant charging and discharging a considerable amount of material is loosened from the plates and settles at the bottom of the cells and this deposit must be removed or it may cause a short circuit inside of a cell unless the battery is designed so that there is enough space at the bottom to hold the accumulation under the plates. The insulators between the plates also have to be replaced occasionally, and there may be one or more broken plates which should be renewed. Perhaps a sulphated condition exists of which one is not aware, and this will come to light when the battery is taken apart and examined.

Effect of Freezing the Battery.—A battery that is well charged will not freeze until a temperature of 96° below zero is reached, and therefore low temperatures are not especially harmful to charged storage batteries. However, when a battery is discharged and the specific gravity falls to 1,025 on the hydrometer, the electrolyte will freeze at a temperature of 30° above zero. The result of freezing is usually complete ruin to the battery, for the active material on the plates will drop away from them. Fig. 79 shows the effect of freezing.

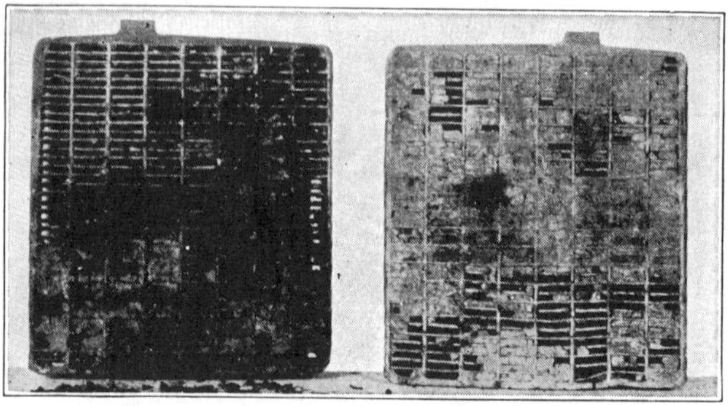

Fig. 79. These Battery Plates Were Ruined by Freezing. When a Battery Is Kept in a Well-Charged Condition There Is no Danger of Freezing the Electrolyte

Overheating the Battery.—Overheating of a battery may be caused by too heavy a rate of charging, by insufficient electrolyte, or by internal short circuits in the cells. The plates become buckled or broken, and this may in turn cause the insulators to break, permitting adjacent plates to touch each other, which also causes a short circuit. Besides warping the plates, overheating may cause the material on the plates to shed and drop away from the grids. Furthermore, overheating causes the electrolyte to get hot, which results in excessive sulphation of plates and also causes the electrolyte to evaporate quickly, resulting in a low level.

Life of Storage Battery.—The life of a storage battery depends on the number of times that it is charged and discharged. If the receiver is used every day for six or eight hours, the battery will have to be recharged much more frequently than a battery of the same size used only one or two hours a day at the same rate of discharge. The thickness of the battery plates determines the number of times that they can be recharged, the positive plates being twice as thick as the negative plates. Storage batteries in autos have an average life of two years, and radio storage batteries, which are not subjected to nearly such hard usage as auto batteries, last from four to eight years with proper care.

Idle Storage Battery.—In case the radio owner is going away for a few months, or for any other reason does not intend to use the battery for some time, he should take it to a battery-service station, where it will be properly taken care of during his absence. Here the battery is given a freshening charge each month to keep it in a fully charged condition, and water is added whenever necessary.

Installing Battery in Basement.—Fig. 80 shows how the batteries and charger can be installed in the basement in cases where it is not desired to have them near the receiver or in a console. Such an installation is particularly advisable when the charger is of the mechanical vibrating type and makes an annoying hum, or if the battery is an old one and is apt to leak. It is best to provide a shelf or platform near the ceiling of the basement directly underneath the position of the receiver, so that the leads from the battery to the receiver will be as short as possible. Here the battery will be out of reach of children, no damage will result from a little electrolyte running over, and a charger hum is permissible. Multi-wire wall plugs, with cables attached, also shown in Fig. 80, can be obtained if it is desired to run the leads to the basement inside of the wall. Such an installation makes a neat apearance, and there is less danger of breaking the leads, as they are not exposed.

Using Battery of Farm-Lighting Plant.—Many farmers have a 32-volt electric-lighting plant, which has a storage battery of 16 cells. Three of these cells can be used to provide the filament current for the radio receiver, provided different sets of cells are used successively so that the drain will be equally distributed over all of the cells. Fig. 81 shows how to obtain 6-volt current for the receiver from three cells of the lighting plant. The leads are attached to spring battery clips, so that it will be an easy task to make connections. Both leads are connected to a double-pole, single-throw knife switch as shown, which must be pulled out every time that the battery is being charged to eliminate the danger of burning out the tubes of the receiver. It is well to have one filament lead distinguished from the other, so that the positive and negative terminals of the receiver will not be reversed in polarity when the leads are connected to the battery.

Tabs of the kind shown in the detail, indicating the polarity of each wire, can be used for this purpose, or a red-colored wire may be used to carry the positive current.

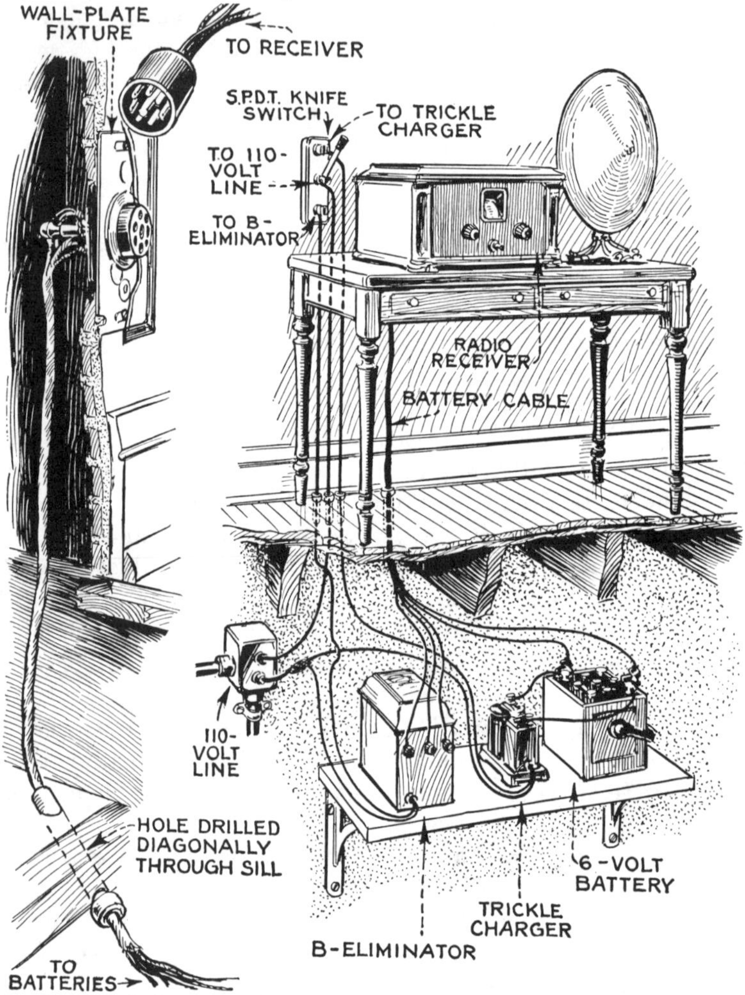

Fig. 80. Method of Installing a Battery Charger and Eliminator on a Shelf in the Basement, a Multi-Wire Cable Being Used to Make Connection with the Receiver. Left: Concealed Installation of Wires in Partition

Chargers

Necessity for Chargers.—To renew the charge of a storage battery, a device must be used which will produce a direct current at a voltage higher than that of the battery, so that the current will be forced through the battery. Most radio owners in cities have 110-volt alternating or direct current in their homes, which can be used for charging the storage battery. However, alternating current must first be rectified to direct current before it can be used for charging purposes. This is done by means of rectifiers, and those used for battery charging are commonly called chargers. In case direct current is used for charging, a charger does not rectify the current but merely controls the current flow into the battery.

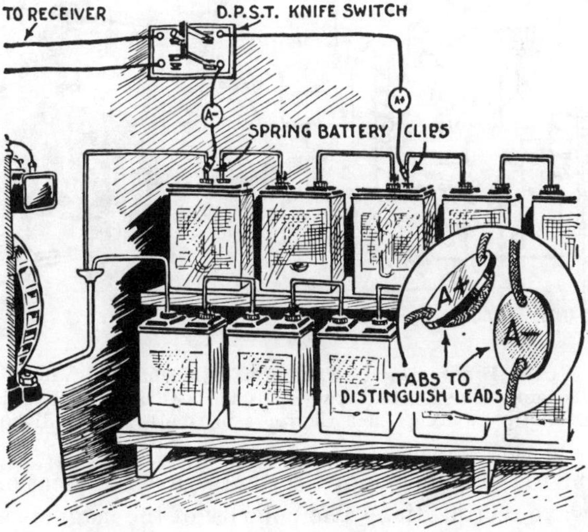

Fig. 81. A-Supply for a Receiver on the Farm Obtained from Farm-Lighting Plant by Tapping Three Cells of the Storage Battery

Chargers are not expensive, and they are really a necessity for the average radio fan. Of course, he can take his battery to a service station to have it recharged, but as a storage battery is rather heavy a few trips to the service station will soon convince any radio owner that he needs a charger. Besides the inconvenience, the cost of having batteries recharged at a service station a few times will equal the cost of a charger, and therefore it is really cheaper to purchase a charger at the beginning.

There are a number of chargers on the market costing from \$2.00 to \$20.00, depending on the size and quality. There are mechanical vibrating chargers, electrolytic chargers, metallic chargers, and gas-filled bulb chargers. Those delivering only a small amount of current at the rate of about ½ ampere or less are called trickle chargers.

Mechanical Vibrating Chargers.—A mechanical vibrating charger is shown in Fig. 82. It consists of a step-down transformer and a vibrating mechanism. The latter comprises one or two electromagnets into which the secondary current from the

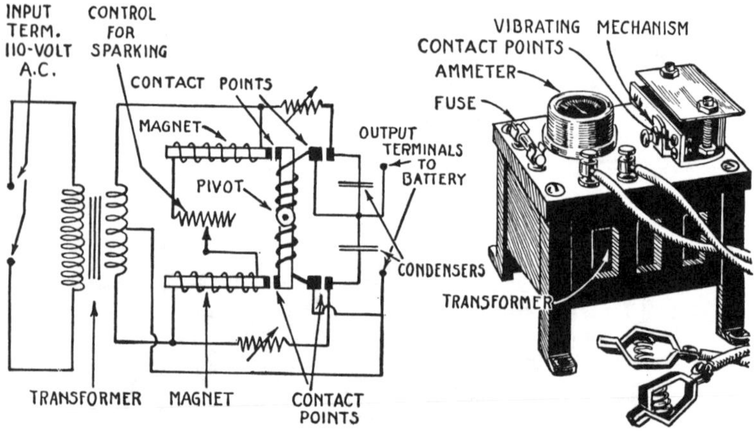

Fig. 82. Wiring Diagram and Drawing of a Mechanical Vibrating Type of Battery Charger. The Contact Points Must Be Cleaned and Adjusted Occasionally to Keep Such Chargers in Operating Condition

transformer is fed, one magnet for half-wave rectification and two for full-wave rectification. The polarity of the magnets is reversed in synchronism with the alternating-current reversals, and they alternately attract and repel a rocking armature. The armature has two contact points which make alternate connection with a pair of stationary contact points so that only one side of the alternating-current wave passes into the output lines, resulting in direct current. Usually it does not make any difference to which terminals of the battery the charger leads are connected, as the polarity of the charger is determined by the polarity of the battery. It has been found best practice with such chargers to reverse the connections to the battery each time that the charger is used as this seems to prolong the life of the contact points.

One of the troubles often experienced with this type of charger is sparking contact points. This trouble may be caused by differences in the alternating-current wave or by the breakdown of the condensers shunted across the contact points. Sparking contact points are likely to become pitted and burned, which sets up a high resistance and tends to decrease the output current. Occasional adjustment of the contacts is necessary, and sometimes they must be filed clean with a small flat file which can be obtained at any hardware or auto-supply store. The contact screws are usually locked in place with small setscrews and these must be loosened before the contact screws can be adjusted. After the adjustment has been made the setscrews are again driven down tight. Mechanical vibrating chargers are sturdy and quite dependable but they produce a hum which is more or less objectionable. However they are more economical to operate than any other kind of charger, although they must be looked after more frequently.

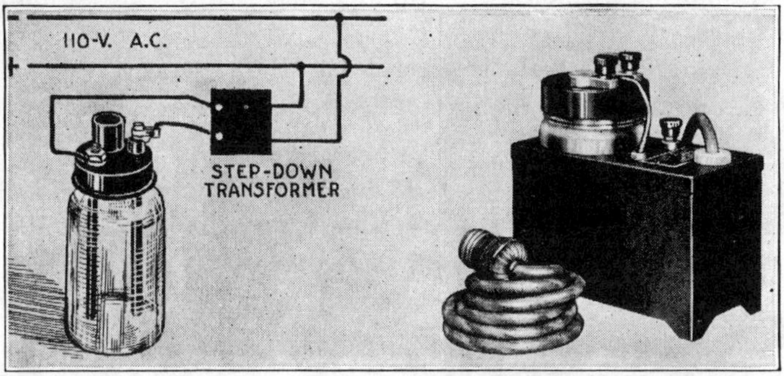

Fig. 83. Left: "Forming" an Electrolytic Rectifier with the Aid of a Step-Down Transformer Connected to the 110-Volt Line. Right: A Well-Made Trickle Charger Using an Electrolytic Rectifier

Electrolytic Chargers.—An electrolytic charger is shown in Fig. 83. It consists of two electrodes partly immersed in an electrolyte. The applied current causes the electrolyte to deposit a film on one of the electrodes, and this permits the passage of an electrical current in one direction only. One of the electrodes is aluminum or tantalum and the other is lead or polished steel. With some types of electrolytic chargers the electrolyte becomes excessively hot and may even boil over, or it may crack the jar that holds it. The water contained in the electrolyte is subject to evaporation and must, of course, be replaced in order to maintain the correct level, as specified on the manufacturer's instruction sheet.

Makers of these rectifiers have tried various methods of overcoming evaporation losses, and some even provide a layer of oil on the electrolyte.

The electrodes of electrolytic chargers are subject to decomposition and must be renewed occasionally. This trouble is worse with some types of electrolytic chargers than it is with others. Impurities in the solution often cause the electrodes, especially the aluminum one, to "cake" up considerably. When the radio owner has to mix the electrolyte himself he should be sure to get the pure ingredients; for instance, where a solution of borax and water is used, ordinary household borax, which may contain some impurities, should not be used but only the chemically pure borax obtained at a drug store. Excessive heating of the electrolyte under a normal load indicates that the solutions and electrodes need attention. An alternating-current hum or buzz indicates a weak solution which will rapidly eat away the aluminum electrodes.

"Forming" Electrolytic Rectifiers.—Before electrolytic rectifiers can be used, the aluminum electrode must be covered with a

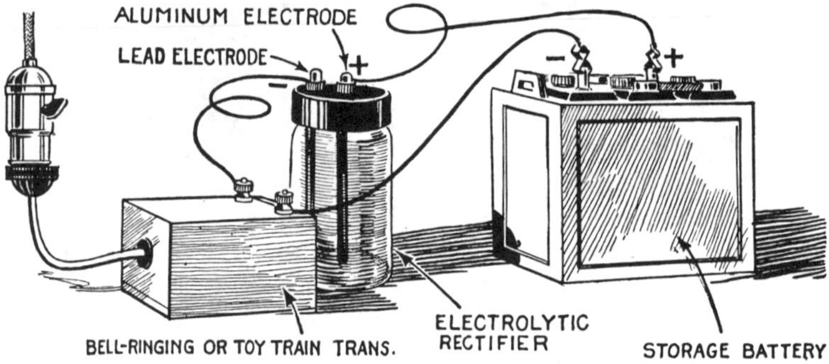

Fig. 84. Method of Connecting a Trickle Charger of the Electrolytic Type to a Storage Battery and to the 8 or 14-Volt Output of a Bell-Ringing or Toy Train Transformer

film of hydroxide of aluminum; the formation of this film is called "forming." Usually the manufacturer takes care of this detail, especially if the rectifier is completely assembled, but in many cases the radio owner must "form" the rectifier himself. This is done by connecting the rectifier to the output of a bell-ringing or toy-train transformer, which is in turn connected to the house-lighting circuit, as shown in Fig. 83, left detail. After the aluminum plate turns to a light gray color, resulting from the coating of oxide on it, it is formed, and ready for use. The charger is then connected in series with the storage battery and the transformers, as shown in Fig. 84. The aluminum plate must always be connected to the positive terminal of the storage battery, while the lead plate is connected to the negative terminal of the battery. If the leads are reversed from this arrangement the charger will discharge the battery instead of charging it. When an electrolytic charger is not in use for a long period of time, the aluminum electrode will gradually lose its film of hydroxide of aluminum, and must then be re-formed. To prevent the loss of the film during a long inoperative condition of the charger, remove the electrodes from the solution and wipe them dry.

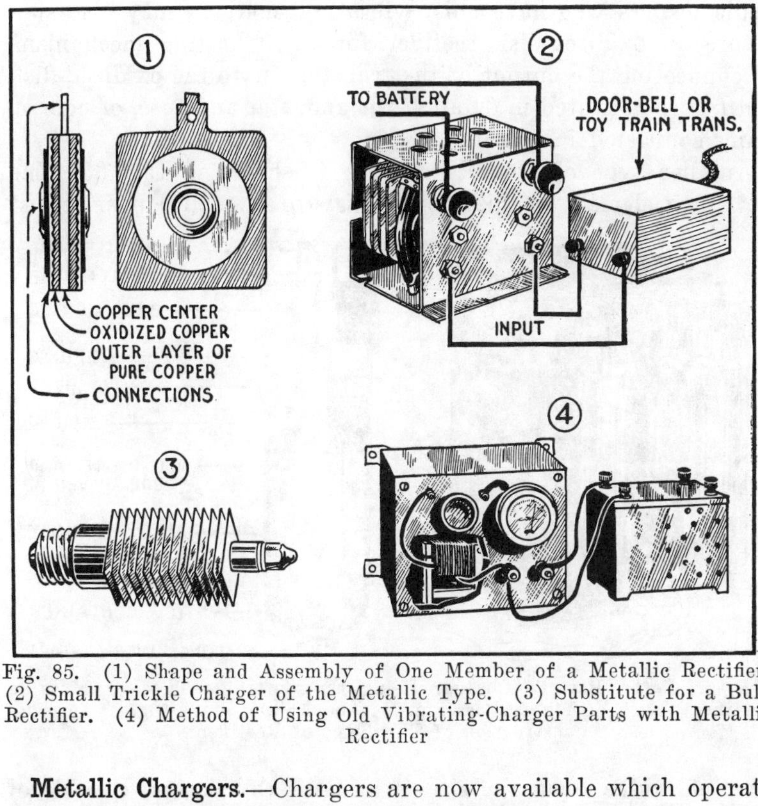

Fig. 85. (1) Shape and Assembly of One Member of a Metallic Rectifier. (2) Small Trickle Charger of the Metallic Type. (3) Substitute for a Bulb Rectifier. (4) Method of Using Old Vibrating-Charger Parts with Metallic Rectifier

Metallic Chargers.—Chargers are now available which operate on the metallic principle. One commercial form of this type of charger consists of a number of copper disks, which are covered on both sides with a coating of copper oxide, which in turn is covered with a film of pure copper, as shown in detail 1 of Fig. 85. When the center and outside of such a disk are connected to alternating-current lines the current will pass through the oxide in one direction only, which makes it a rectifier. These chargers can be had in various capacities, the amount of direct current passed being determined by the number of disks that the charger contains. The charger shown in detail 2 is of the trickle type and is connected to a step-down door-bell or toy-train transformer. These chargers can be purchased complete, including the step-down transformer, or the rectifying element only can be had separately, this being manufactured with attachments which permit it to be substituted for a bulb rectifier, as shown in detail 3. Radio owners who have noisy vibrating chargers may also substitute an oxidized-disk rectifier for the vibrating mechanism, by connecting the output of the transformer to the oxidized-disk rectifier as indicated in detail 4, the ammeter and fuse, of course, being connected in one line.

Another type of metallic charger is shown in Fig. 86. The rectifying element in this case consists of a small copper-coated

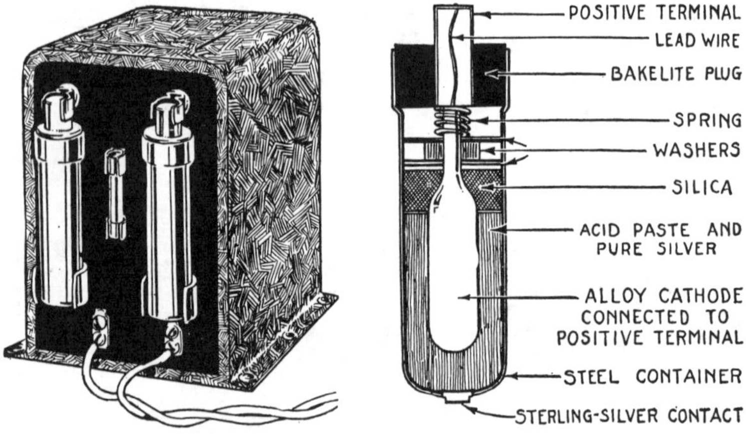

Fig. 86. Left: Battery Charger Using the Cartridge Type of Metallic Rectifiers, Each of Which Passes a 2½-Ampere Current. Right: Internal Construction of This Type of Rectifier

steel cartridge about the size of a man's thumb, and capable of passing 2½ amperes of current. The internal construction of the cartridge is shown in the detail. Current will flow in one direction only, from the positive alloy cathode to the steel shell, which is negative, and makes connection with an external clip through the sterling silver contact at the end. Metallic chargers do not overheat, and much can be said in their favor, especially that they are noiseless in operation and do not have to be looked after to keep them in operating condition. The cartridge type must be renewed occasionally, as it is subject to depreciation.

Gas-Filled Bulb Chargers.—A charger of the kind shown in Fig. 87 has a filament-lighted bulb, which passes from 2 to 5 amperes of current, depending on the size of the bulb. A step-

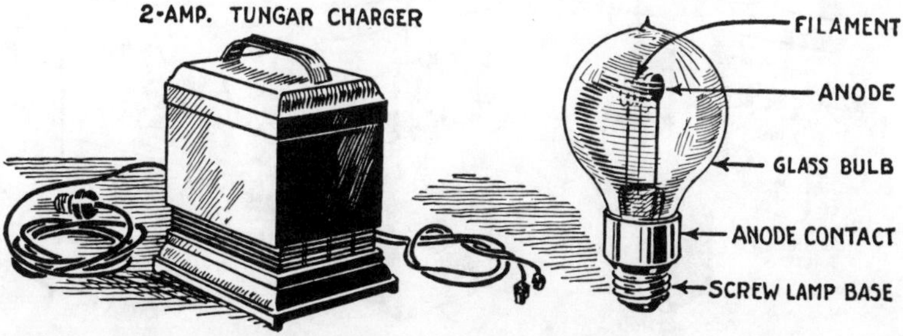

Fig. 87. Battery Charger Which Makes Use of a Filament-Lighted, Gas-Filled Rectifying Bulb. It Is Silent in Operation and Is Capable of Delivering a 2-Ampere Current

down transformer is incorporated with the charger to supply a suitable current for lighting the filament. Ordinarily a slight transformer hum is noticeable with this type of charger. When the switch is closed to connect the storage battery to the charger the transformer hum will drop in pitch. If this is not the case, the bulb may be defective, there may be an open circuit in the transformer, or in the output line of the charger, or there may be corrosion on the battery terminals, which may prevent the charger current from passing into the battery. Rectifying bulbs of this type may be defective, even though the filament continues to light.

Direct-Current Charging Hookup.—In localities where direct current is available, a radio battery can be charged directly from the house-lighting current, a bank of lamps, an electric iron or heater element being connected in series with the battery to control the amount of current passing through it. Such hookups are shown in Fig. 88. To obtain a low-rate or trickle charge, use one

40-watt lamp in series with the battery. This will pass a 2/5-ampere current. Heavier current rates can be obtained by using a lamp of larger size or by substituting an electric flatiron or a heater element of the screw-plug type, which passes from 2 to 5 amperes, according to its size. When using a heater element protective fuses of 10-ampere capacity should be used. If it is not desired to use a large lamp, such as one of 100-watt capacity, or larger, several smaller lamps of the 40 or 60-watt size can be connected in parallel with the first socket, as indicated in detail 2.

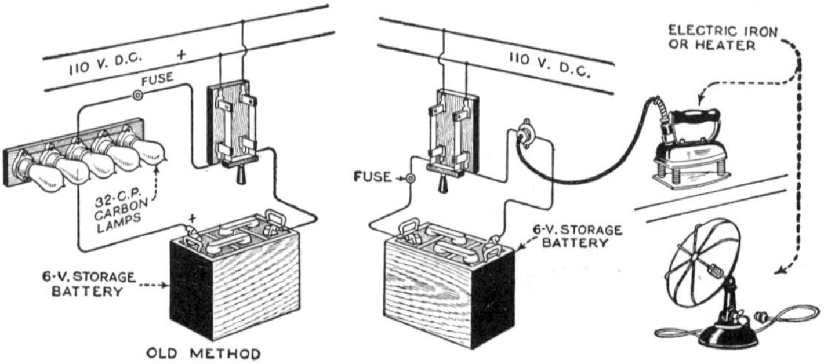

Fig. 88. Left: Method of Using a Bank of Lamps in Series with a Battery to Control the Charging Rate when Connected to 110-Volt Direct-Current Lines. Right: Electric Iron or Heater Substituted for a Bank of Lamps

(Courtesy of "Popular Mechanics Magazine")

It is advisable to use an ammeter on the charging panel to indicate the amount of current being passed. When connecting the leads of the charger to the storage battery, their polarity should be definitely ascertained. This can readily be determined by the salt-water or raw-potato method shown in Fig. 68. Adequate fuses should also be provided as a protection against short circuits. When a battery is discharged, the initial charge may be as high as 8 or 10 amperes, but by substituting greater resistances, such as smaller lamps, a lower rate of charge can be given when the charge is nearly completed. Instead of using lamps and heater elements as just described, a variable rheostat suitable for 110-volt current may be used, if obtainable; this can be adjusted conveniently to pass any desired amount of current.

Protecting the Receiver from Charger Voltage.—When charging a radio storage battery, one should never let the receiver remain connected to the battery, for all chargers produce a potential higher than 6 volts. It has often been the sad experience of many radio fans to forget to remove the battery leads to the receiver before switching on the charger, with the result that several or all of the tubes were burned out. A convenient method of connecting the battery to charger and the receiver is shown in Fig. 89. A double-pole double-throw knife switch is obtained, and the receiver, battery and charger are wired to it as indicated. The moment the blades of the switch are thrown over in one direc-

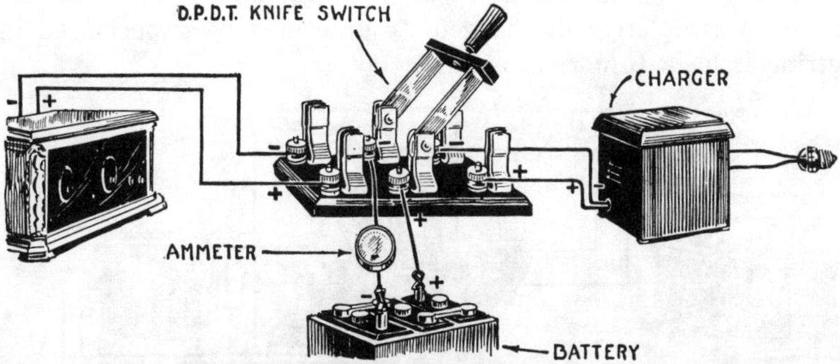

Fig. 89. Convenient Charging Hookup Using a Double-Pole, Double-Throw Knife Switch, Which Eliminates the Possibility of Applying the Charger Output Directly to the Receiver

tion, the receiver is disconnected and the charger is connected to the battery, and both can never be connected to the receiver at the same time. The two leads from the battery are connected to the center terminals of the switch, the charger leads to the terminals on one end of the switch and the receiver leads to the terminals on the other end. When making these connections care should be taken to make sure that the three terminals on each side of the switch are connected to leads of the same polarity, as clearly indicated in the drawing. If desired, the installation can be made still more convenient by using a triple-pole, double-throw knife switch, as shown in Fig. 90, connecting one of the 110-volt input lines of the charger to two of the terminals as indicated. One movement will connect the charger output lines to the battery, and also switch on the charger.

Charger Connections to Battery Reversed.—Only some chargers of the mechanical vibrating type can be connected to the battery without distinguishing carefully between the polarity of the leads and connecting them to terminals of the battery of the same polarity. The output leads of a charger are usually carefully marked, but in case no distinguishing marks indicating the polarity are found, it may be discovered that one lead has a red-colored insulation, and this is then the positive lead. While serious damage will not result if the leads from the charger have accidentally been attached to the opposite terminals on the battery and the charger has been run for a short time while connected in this way, the result of this arrangement is that the battery will be discharged rapidly instead of being charged, and after a battery has been completely discharged, trouble may be experienced in getting it back to normal again.

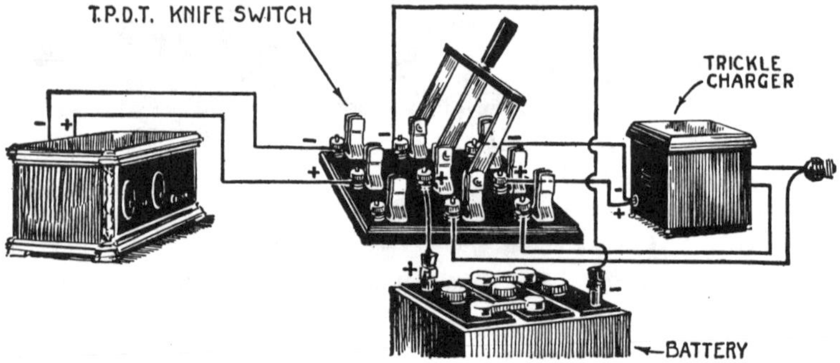

Fig. 90. Charging Hookup Which Turns off the Charger, Disconnects It from the Storage Battery, and Connects the Battery to the Receiver in One Operation, Thereby Eliminating Charger Hum during Reception

How to Determine Duration of Charge.—When a receiver is used more or less regularly and it is possible to determine an average length of time that the receiver is in daily use, it can also be determined approximately how much electricity is taken from the storage battery every day, and how long it will take to replace it. The factors to be taken into consideration are the total current draw of the receiver, the average number of hours that the receiver is used, and the current output of the charger. For instance, a receiver using six UX 201A tubes, which pass .25 ampere each, has a total current draw of 1.5 amperes. If the receiver is used on the average of four hours each day, there will be a daily consumption of 6 ampere-hours of electrical energy. If the charger has a rate of 2 amperes, it will take theoretically, 3 hours to replace the 6 ampere-hour energy which was consumed. However, it must be taken into consideration that a battery on charge offers a higher resistance toward the end of the charge than it does at the beginning of the charge, and this increased resistance decreases the current flow from the charger. Furthermore, allowance must also be made for losses outside of the current consumption by the tubes. Besides, a slight overcharge, now and then, is beneficial. One should therefore figure that the charger should be operated about 20 per cent longer than the time estimated above. Thus in this case it is found that 36 minutes is 20 per cent of 3 hours, and therefore the charger should be run approximately 3½ hours. The loss percentage has here been based on an arbitrary estimate, and it is best to check up on the battery with the hydrometer in order to determine exactly how much time is necessary for the average charge. A daily recharge, to make up for the energy consumed the previous day, keeps the battery up all the time, so that the owner will have the most efficient operation of his receiver and need not be concerned about the possibility of finding his battery discharged. After the daily charging rate has been determined as accurately as possible, it is unnecessary to take a hydrometer reading every day, but this should be done occasionally to check up on the battery condition. However, regular attention must be given to the level of the battery electrolyte.

Trickle-Charger Combinations.—The advantage of daily charging to keep the battery in a fully charged condition has been met by trickle chargers. These chargers give a low rate of charge, which in many cases is adjustable, so that, instead of using them for a certain length of time every day to recharge the battery, they can be left on continuously. It is possible to let them operate even while the receiver is being used, but this practice often results in an alternating-current hum, which is more or less objectionable and it is better therefore to disconnect the charger during reception. It is a good plan to install a switch arrangement, as shown in Fig. 90, which will turn off the charger when the filament current of the receiver is turned on. Some manufacturers have combined a small battery, a trickle charger, and a B-eliminator in a compact unit. One of these units, which has been quite popular, and gives satisfactory results is shown in Fig. 91. The charging rate of the trickle charger contained in this unit is adjustable and can be set at 0.2, 0.33, and 0.6 ampere, according to the current draw of the receiver. A table showing what rate to use is given in Fig. 92. The charger is automatically cut out when the receiver is turned on by means of a switch on the unit, the receiver switch being left on all the time.

Fig. 91. Power Unit Which Consists of a B-Eliminator, a Small 6-Volt Storage Battery, and a Trickle Charger
(Courtesy of "Popular Mechanics Magazine")

When using a trickle charger it is a good idea to determine by means of an ammeter exactly how much current is being drawn out of the battery, and then the charging rate can be adjusted

Filament Current, Amperes	Rate and Hours Service		
	0.2 Amp.	0.33 Amp.	0.6 Amp.
0.25	10.0	13.1	16.4
0.50	6.35	9.00	12.4
0.75	4.65	6.86	10.0
1.00	3.66	5.54	8.40
1.25	3.02	4.64	7.24
1.50	2.57	4.00	6.35
1.75	2.24	3.50	5.56
2.00	2.00	3.13	5.10

Fig. 92. Three Charging Rates at Which the Charger in the "Philco" Power Unit Can Be Set, According to the Amount of Current Drawn and Length of Time a Receiver Is in Daily Use

accordingly. Fig. 89 shows how to hook up the ammeter for this purpose. It is cut in one of the battery leads to the switch.

Troubles Experienced with Trickle Chargers.—Most of the troubles experienced with trickle chargers can be attributed to poor adjustment of the charging rate, which is usually the fault of the owner. The result of poor adjustment of a trickle charger is either an overcharge or an undercharge of the battery, either of which has injurious effects. As a trickle charger is left connected to the lighting circuit all the time, there is danger of forgetting to shut it off when the receiver is not being used for a long time. This also results in a prolonged overcharge, which may damage the battery. Trickle chargers of the electrolytic type have to be inspected now and then to see whether or not the level of the solution has dropped, and if so, distilled water should be added. Too low an electrolyte level will cause undue heating of the charger, and may even ruin the electrodes. The maintenance of the proper level of the electrolyte is very important.

Charging Storage Batteries on the Farm.—Where lighting current is not available, as on the farm, other methods must be used for charging the storage battery. There are at present about a million farms equipped with radio-receiving sets, and the matter of charging radio batteries on the farm is an important one. Almost every farmer owns an automobile, and in most cases the radio battery can be connected in parallel with the car battery, as shown in Fig. 93. A still better method of connecting a radio

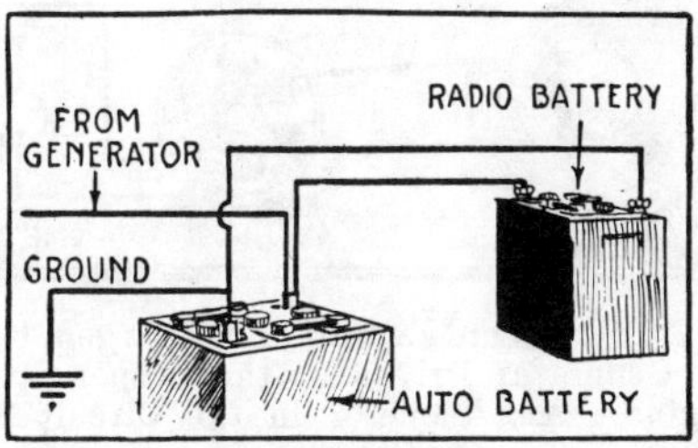

Fig. 93. Recharging a Battery by Connecting It in Parallel with an Auto Battery

battery to the generator circuit of a car is shown in Fig. 94. A few hours' ride will usually charge the battery so that it can be used again. This is more or less of an emergency method, and it is better to purchase a second-hand 6-volt automobile generator, and connect it to a small gasoline engine, as shown in Fig. 95. Connect the outlet of the generator to the storage battery and run the generator just fast enough to produce a 2-ampere current, a small ammeter being cut in one of the leads to the battery to indicate the charging rate. When connecting up the generator to the battery, be sure to find out the polarity of each lead, which must be

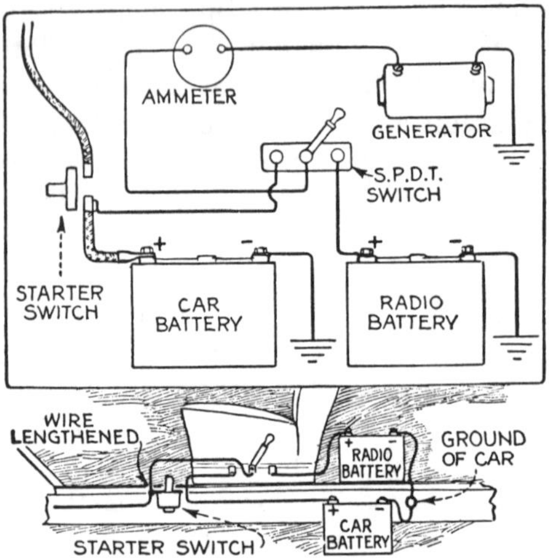

Fig. 94. Using the Auto Generator to Charge a Radio Battery, Cutting out the Regular Auto Battery from the Circuit

(Courtesy of "Popular Mechanics Magazine")

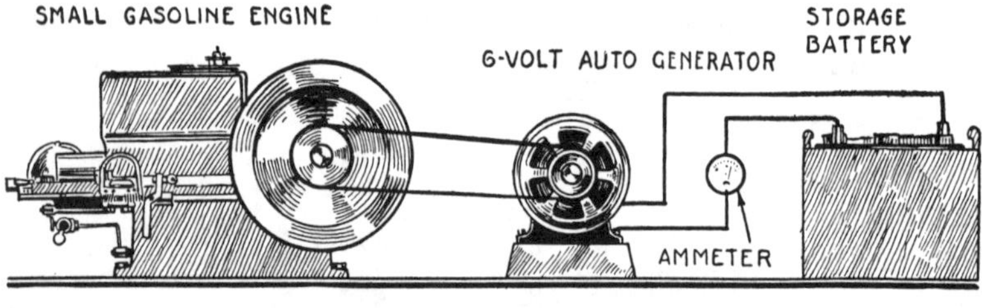

Fig. 95. A Radio Battery Can Readily Be Charged by Means of a Discarded 6-Volt Auto Generator Driven at the Proper Speed by Means of a Small Gasoline Engine. An Ammeter in the Circuit Shows the Rate of Charge

connected to a terminal on the battery of the same polarity. Many generators are grounded, that is, one of the brushes is connected to the shell or case of the generator, leaving only one output terminal. In such cases a good electrical connection must be made to the case or shell, which can be done by drilling and tapping the case with a No. 34 drill and then using a 6-32 tap. The metal around the hole is filed clean, and a short 6-32 screw is driven down to hold the bared end of the lead.

If the farm is equipped with a 32-volt direct-current lighting plant, the battery can be given a charge from any light socket on the circuit, this being done, preferably, while the generator is running. There must be a suitable resistance between the light socket and the battery to control the amount of current passed. Fig. 96 shows how such a charging outfit is arranged. The element

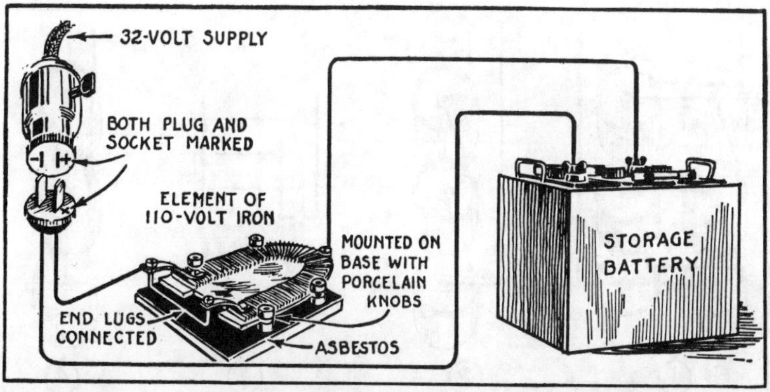

Fig. 96. Charging a Radio Battery from a 32-Volt Farm-Lighting Circuit by Passing the Current Through the Element of a Discarded 110-Volt Iron, Which Controls the Charging Rate

from a discarded 110-volt electric iron, which can be obtained at almost any electrical store, has about the right resistance for this purpose, if connected as shown in the diagram. The two leads which supply the charging current are attached to spring battery clips to facilitate attachment to the terminals. While charging the battery, the ammeter needle should be deflected to show charge. If this is not the case, reverse the plug in the light socket. It is a good idea to mark the plug and socket so that the plug will always be inserted correctly.

Dry-Cell A-Batteries

How to Hook Up Dry Cells.—For receivers equipped with 5-volt tubes, which pass a .25-ampere current each, the use of dry cells for supplying filament current is expensive and therefore not advisable, although they can be used temporarily. But the UX 199 tubes, or similar types of tubes, which are designed to operate on a potential of 3 volts and pass only a .06-ampere current, ordinary dry cells can be used satisfactorily. A battery of either 3 or 6 standard dry cells is used, depending on the size of the receiver. A 6-cell, series-parallel arrangement, as shown in detail 1 of Fig. 97, will last much longer than two separate 3-cell batteries as shown in detail 2. But where weight has to be taken into consideration, as, for instance, in portable sets, the use of a 3-cell battery is advisable. Where it is highly essential to have light

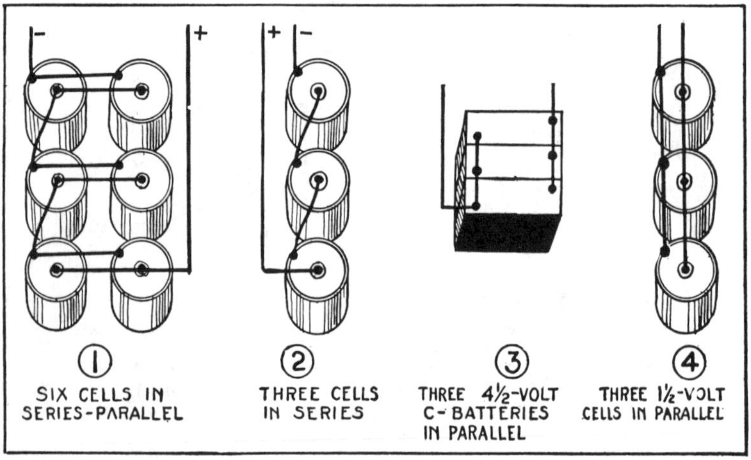

Fig. 97. (1) Connection of Six Dry Cells for 3-Volt Tubes. (2) Three Cells in Series Where Lighter Weight Is Required. (3) Three C-Batteries Where Minimum Weight Is Essential. (4) Three Dry Cells in Parallel for 1½-Volt Tubes

weight a few 4½-volt C-batteries hooked up in parallel, as shown in detail 3, can be used on a one or two-tube receiver. WD 11 or C 11, and WD 12 and CX 12 are 1.5-volt tubes and need a battery consisting of a few dry cells hooked up in parallel, as shown in detail 4 of Fig. 97.

Life of Dry-Cell A-Batteries.—The length of time that a dry-cell battery will give service depends on the current draw of the receiver, on the number of hours that the receiver is used, and also on the construction and quality of the dry cells. Fig. 98 gives a table prepared by a well-known battery manufacturer, which gives the approximate hours of service obtainable from the highest quality dry-cell A-batteries. When they have passed their period of usefulness, and are no longer suitable for radio purposes they can still be used for ignition and door-bell service if desired. The common type of dry cell is not rechargeable when once exhausted, although it is true that dry cells recuperate to some extent when left standing for a while. However, this recuperated condition is only temporary, and a test of dry cells after they have been left standing unused for some time is not a true indication of their condition. When completely discharged, dry cells are likely to swell, and a moist paste or pulp often seeps out, especially at the seams. This substance may cause stains on the inside of the receiver cabinet where the cells have been standing, or the connecting wires may become corroded, for the pulp contains a certain percentage of acid, which will attack metal. Dry cells should be discarded long before they reach this condition.

Type of Tube	UV or UX-199 C or CX-299 DV-3		WD-11 or C-11 WD-12 or CX-12		UV-201A C-301A DV-2		UV-200 C-300
For	Tube Amps.	Hrs.	Tube Amps.	Hrs.	Tube Amps.	Hrs.	
1 Tube............	0.06	700	0.25	110	0.25	110	Not a Dry-Battery Tube
2 "	0.12	300	0.50	110	0.50	110	
3 "	0.18	180	0.75	110	0.75	110	
4 "	0.24	150	1.00	110	1.00	110	
5 "	0.30	240	1.25	110	1.25	110	
6 "	0.36	180	1.50	110	1.50	110	

Fig. 98. Table Showing the Average Life of Good Dry Cells when Used for Supplying A-Current to a Radio Receiver

Testing Dry Cells.—When dry cells are purchased, the dealer usually tests them by connecting a 0 to 30-reading ammeter across their terminals. The ammeter is short-circuited across the dry cell momentarily, as shown in Fig. 99, and this method of testing is therefore called "flashing." Such a test indicates only the short-circuited amperage, which should be from 22 to 30 amperes on a new cell. When a cell gives a reading of 5 amperes it can no longer be used for radio purposes. However, such a reading is not always a true indication of the actual condition of the cell.

A much more accurate method of testing dry cells is to use a low-reading voltmeter, graduated in tenths of a volt, as also shown in Fig. 99. A new dry cell should test 1.5 volts when its terminals are not connected to a receiver or to any circuit whatever. As soon as the voltmeter indicates only 1.2 volts, the cell is no longer suitable for supplying filament current and should be discarded. Usually all the dry cells of a battery will be found to be in the same condition.

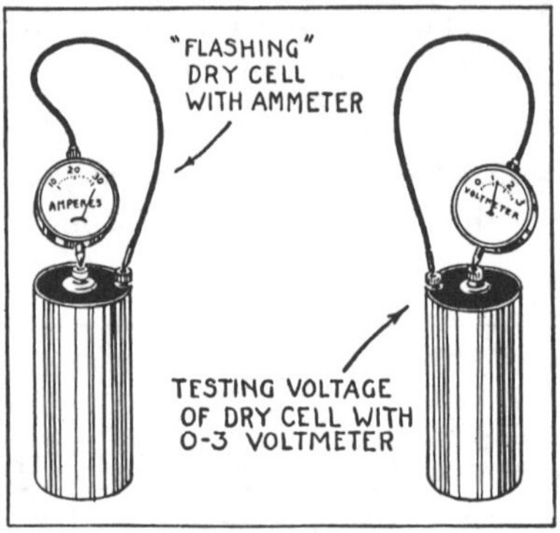

Fig. 99. Dry Cells Can Be "Flashed" with an Ammeter but Can Be Checked More Accurately by Means of a 0-3 Voltmeter

Results of Low Dry-Cell A-Battery.—When an A-battery becomes low, the receiver will tune broad, and it will be found necessary to advance the filament rheostats almost entirely over to the right, supplying all the available current to the tubes in order to get the same degree of volume that was obtained when the rheostats were turned up only one-fourth or one-third as high at the time when the A-battery was new. No attempt should be made to rejuvenate dry cells by the addition of certain solutions, by heating them, or by any other method, for this is merely a waste of time. When they have passed their useful stage they should be discarded and new ones should be purchased. When replacing worn-out dry cells care should always be taken to turn off the receiver and remove the tubes. After the new cells have been connected up in the proper way, all the rheostats should be turned down before turning on the switch. These precautions are necessary to avoid burning out the tubes in case of an accidental short circuit while connecting the batteries, and the rheostats are retarded to prevent too high a filament current from passing through the tubes, which might paralyze them.

B-Batteries and Their Troubles

B-batteries are used to provide plate current for the tubes of a radio receiver. They furnish a steady direct current, and they have been perfected to such an extent that they will last a long time, their life depending, of course, on their size and on the amount of plate current consumed by the receiver. They are, however, being superseded by B-eliminators, but there are many places, as for instance on the farm, where electric-lighting current is not available for operating B-eliminators, and then B-batteries must be resorted to.

Internal Construction of B-Batteries.—B-batteries consist of a number of small cells, similar to flashlight cells, which are connected in series, that is, the positive terminal of one cell is connected to the negative terminal of the next, and the positive terminal of the second to the negative terminal of the third, and so on, leaving a positive terminal at one end of the battery and

Fig. 100. Internal Construction of a B-Battery Showing How the Small Cells Are Arranged and Connected

a negative terminal at the other end. These latter are connected to the two lugs or screws which project above the top of the battery. A center tap is also usually provided and connected to an outside terminal. Each small cell is capable of producing a pressure of 1.5 volts when the battery is new, and the total voltage of the battery is equal to the number of cells multiplied by the voltage of one cell. Thus, a 22½-volt battery has 15 cells, and a 45-volt battery has 30 cells. The cells are small in order to minimize the weight and size, for large, bulky batteries are undesirable, and small batteries are especially suitable for portable receivers. The cells are usually arranged in tiers, and are set in a cardboard or metal box. The cells are insulated from each other by a compound, which is poured around them at the time of manufacture and hardens in a short time. The top of the battery usually consists of a heavy layer of sealing wax. The internal construction of a dry-cell B-battery is shown in Fig. 100.

A number of standard dry cells, connected up in series, would also serve the same purpose, but the average radio owner has no space for a battery of such large dimensions. For instance, on a receiver which requires a plate potential of 180 volts, it would be necessary to supply 120 dry cells, and this would, in most cases, be impracticable. However, the plate current of a receiver is so small, being measured in milliamperes, or 1/1000th parts of an ampere, that a battery of small cells is satisfactory, although the standard cells would last much longer.

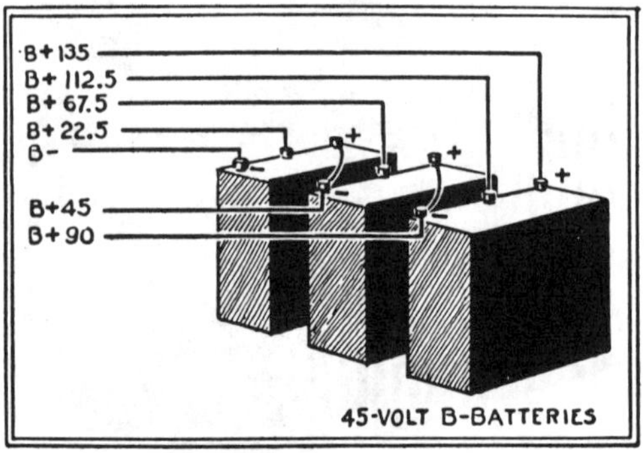

Fig. 101. Three 45-Volt B-Batteries Connected in Series Showing Various Taps Which Give from 22.5 to 135 Volts

How to Hook Up B-Batteries Properly.—Most one-tube receivers require only 22½ volts of plate potential, and only one B-battery is then required. Others require 45, 90, 135, or 180 volts of plate voltage, depending on the number and type of tubes in the receiver and its method of audio coupling. Then three or four 45-volt B-batteries are required, and they are hooked up in series as shown in Fig. 101, taps being taken off at various terminals for radio-frequency, detector, and audio-frequency tubes according to the needs of the receiver. Usually a voltage of 45 to 90 volts is necessary for the radio-frequency tubes, and a voltage of 22½ to 45 volts for the detector. In transformer-coupled receivers which do not have a power tube, 90 volts of plate potential are required, and in resistance and impedance-coupled sets, 135 volts are needed for the audio-frequency tubes. Where a power tube is used, a higher voltage is necessary on this tube. Information concerning the proper plate potentials for such tubes is usually given on the manufacturer's instruction sheet, which accompanies the receiver. Also see Fig. 153.

Testing B-Batteries.—Although in a few cases dealers resort to an amperage test on B-batteries, such a test means very little and

Fig. 102. B-Batteries Should Be Tested Occasionally with a Suitable Voltmeter to Insure Their Being in Good Condition
(Courtesy of "Popular Mechanics Magazine")

does not show their true condition. The only accurate test on B-batteries is a voltage test, a 0 to 50-reading voltmeter being used for this purpose, each cell being tested separately, as shown in Fig. 102. B-batteries should never be tested for voltage after they have been standing idle for some time, for when standing idle, they recuperate to some extent and during the first few moments of their use they will produce a higher voltage than they will after being in use for several minutes. Therefore a voltmeter test should be made after the receiver has been operated for some time.

Life of B-Batteries.—When a 22½-volt battery drops down to 17 volts, or a 45-volt battery drops to 37 volts, its efficient service for radio purposes is over, and it should then be replaced by a new one. It is true, as stated by a well-known manufacturer of B-batteries, that a 22½-volt battery, which has dropped to 17 volts, can be connected up with other old ones to supply voltage

Size of B-Battery in pounds	Current Draw in Milliamperes						
	2	5	8	10	15	20	30
1..........	350	100	....	....	....	....	....
2..........	900	300	140	110	....	....	....
5..........	1,800	1,000	500	400	230	150	75
7..........	3,600	2,000	1,250	1,000	600	400	215
	Service Life in Hours						

Fig. 103. Table Showing the Average Life of Good Radio B-Batteries of Various Sizes and Subjected to Various Rates of Discharge

(Courtesy of Burgess Battery Co.)

for the amplifier tubes, until its voltage drops down as low as 10 volts, but most batteries are inclined to become noisy when they have reached the 17-volt point, and it is therefore unwise to use them any longer. Fig. 103 shows a table which contains the figures of a large battery manufacturer concerning the life of B-batteries of various sizes and under various loads of plate-current draw.

Determining Amount of Plate-Current Draw.—The amount of plate current passed by a receiver can be determined with the use of a 0 to 50-reading milliammeter, connected in the negative line from the B-battery to the receiver, as shown in Fig. 104. This gives the total amount of plate current passed by the receiver. To determine the plate current of the detector separately, or that of the radio-frequency or audio-frequency tubes, the milliammeter is connected in the positive B-supply line which provides plate current for the tube to be tested.

The total plate-current draw of a receiver depends on the type and number of tubes it has, on the grid bias, and on the kind of coupling used between the tubes. It is generally conceded that a good resistance-coupled receiver will draw less plate current than a transformer-coupled or an impedance-coupled receiver having the same number of tubes. Proper C-biasing on the grid of the audio tubes will help considerably to prolong the life of the B-batteries, for it greatly reduces the drain on them. Besides being useful in determining approximately how many hours of service the B-batteries will give, the milliammeter test of plate-current draw will tell whether a receiver will be expensive in respect to the B-batteries needed for it. It must, however, be remembered

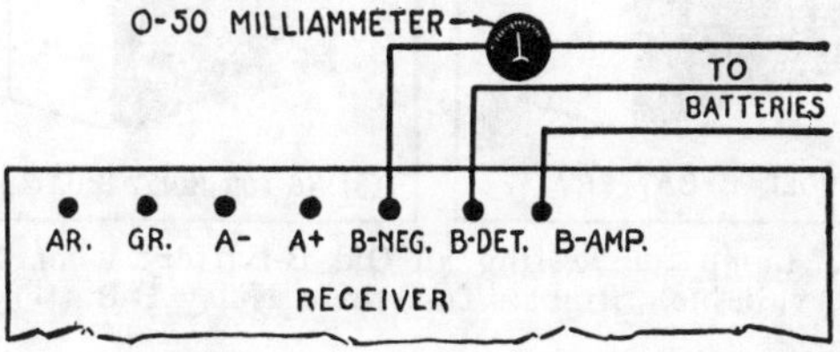

Fig. 104. Method of Determining the Total Plate-Current Draw of a Receiver by Connecting a 0-50 Milliammeter in the B-Negative Line

that fresh B-batteries often give an exceptionally high milliampere reading, which may be about 15 per cent above the normal current used under average working conditions. When a milliammeter is connected up to show the total plate-current draw, the needle should remain more or less steady. Appreciable fluctuations of the needle indicate that either the B or C voltage is incorrect, or possibly both. If the needle drops toward zero, when strong signals are reproduced, the C-voltage is too low, and if the needle rises at a strong signal, the B-voltage is too low.

Exhausted B-Batteries.—Low B-batteries cause weak, distorted, and wavering signals, and distant reception is then entirely impossible. When this symptom manifests itself and the owner is positive that the A-battery is fully charged and that the tubes are all in good condition, the B-batteries are presumably at fault, and they should be tested with a voltmeter as described above. In case the owner has no voltmeter at hand, but suspects exhausted B-batteries, he can readily test them by the simple method shown in the left detail of Fig. 105. A 10 or 25-watt lamp and a length of solid copper wire about 18 in. long should be obtained. The wire is bared at one end for about 6 in., and is wrapped around the screw section of the lamp as indicated. The center contact of the base of the lamp should be held firmly against one terminal of the battery momentarily. If the lamp does

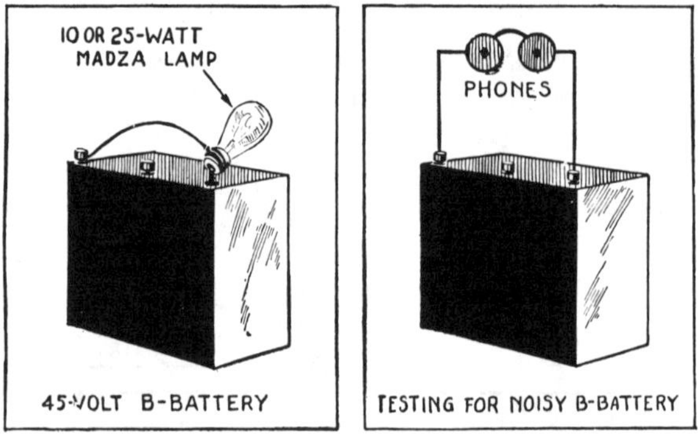

Fig. 105. Left: Lamp for Testing an Old B-Battery when No Voltmeter Is Available. Right: Test for a Noisy B-Battery

not glow at all while this is being done, or if it glows very faintly, the battery is exhausted and should be discarded. This emergency method should never be tried on good batteries, because the lamp draws a considerable current and will quickly drain a battery. A prolonged light of only a few minutes could be obtained on a lamp in this way, and a good battery would then be worthless. This test cannot be applied to 22½-volt batteries.

Noisy B-Batteries.—When 22½-volt B-batteries drop down to 17 volts, and 45-volt batteries to 37 volts, they may become noisy. Even new batteries may sometimes be noisy, which results in very unsatisfactory reception. To find out whether B-batteries are noisy, a pair of headphones should be obtained, and the batteries can be tested as shown in the right detail of Fig. 105. The tips of the phone cord should be screwed down tightly under the terminal nuts of the batteries, as any slight movement will cause a noise in the phones and give an incorrect test. With the tips held down securely, listen for noise. Noise in B-batteries is caused by fluctuations in voltage, resulting from poor or defective cell insulation which is not entirely moisture proof, by the quality of the compound used in the cells, by impurities in this compound, by faulty cell construction, and by loose internal connections. It is, of course, impossible to manufacture perfect batteries, and such old-age noise is really beyond the manufacturer's control, although it is true that well-made batteries containing good material will generally not be so prone to become noisy as cheaper-grade batteries. Sometimes the voltage fluctuations are so violent that they cause considerable deflection of a voltmeter needle, and when connected to a receiver such batteries may cause loud, crashing noises.

Open-Circuited B-Batteries.—Total inaudibility can sometimes be traced to internally open-circuited B-batteries. This may be attributable to one cell being exhausted prematurely by a broken connection, or by a poorly soldered joint between the cell connectors and cell terminals. This trouble may even exist in a new battery used only a few hours, which was still in good condition when

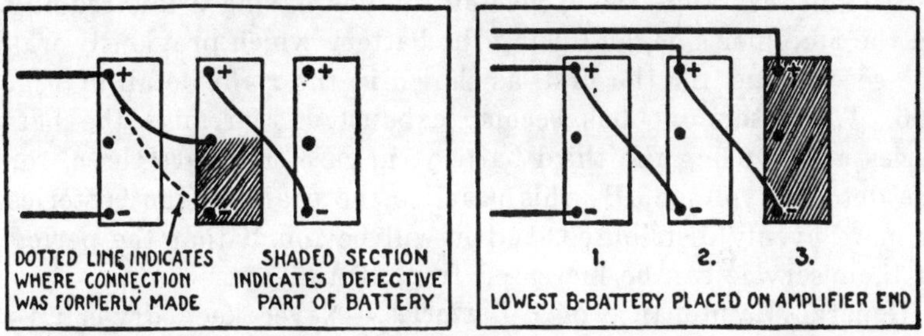

Fig. 106. Left: Shunting out an Open-Circuited Section of a B-Battery. Right: Proper Rotation of B-Batteries Distributes Plate-Current Consumption and Maintains the Efficiency of a Receiver

it was sold and tested by the dealer. A voltmeter is used to locate the trouble: Hold one lead of the voltmeter to the center tap and the other lead to one end terminal of the battery. If the voltmeter needle deflects, the cells between these two points are in good condition. Then touch the center tap and the other terminal in the same way and look for the needle deflection. Proceed in the same way with the other batteries. As soon as two terminals are touched and the needle fails to move, the open circuit is located. This test can be made while the batteries are connected to the receiver, but it is advisable to remove the tubes as one might accidentally touch a filament and a B-positive wire or terminal at the same moment, which might burn out the tubes. In case the battery was purchased shortly before the trouble developed, it should be returned to the dealer and he will perhaps replace it. If the battery has been in use for some time and no replacement is made, the good side of the battery can still be used, provided the voltage of this part has not dropped too low for satisfactory use. Fig. 106, left detail, shows how the defective part is left out of the circuit by connecting the other batteries to the two "live" terminals of the faulty battery. If the negative terminal of the battery is on the defective side, then the center tap is used as the negative terminal.

Rotation of B-Batteries.—The battery used to supply detector plate voltage is usually the first to become exhausted, because of the greater plate-current draw of the detector circuit in a receiver not equipped with a power tube. This battery or section of a battery will drop down to the low-voltage limit before the other batteries, and therefore it is a good idea to disconnect the batteries and rearrange them, placing the one having a low reading on the amplifier side, and using the battery which previously was the second one, for the first, as shown in the right detail of Fig. 106. When this one has become exhausted, rearrange the batteries again, using the third battery, in case there are three, for the detector voltage. By this practice the drain on the batteries is more evenly distributed, and it will be found that the period of their service can be increased from 10 to 20 per cent.

General Maintenance of B-Batteries.—Never keep dry-cell B-batteries in places where they are exposed to extremes of temperature, as, for instance, on window sills during cold weather, or near radiators, for such exposure reduces their life considerably. Moisture is also more or less injurious to dry-cell B-batteries, for if it gets inside and saturates the insulation between the cells, there may be a slight electrical leakage and noisy batteries may be the result. B-batteries are left connected to the receiver all the time. The idea that current is constantly being drawn out if they are left connected while the receiver is not in use is erroneous, as plate current is passed through the tubes only when the filament is heated. The tops of the B-batteries should be kept free from dust, which can readily be done by brushing them off occasionally, as shown in Fig. 107. If this is not done, moisture from the atmosphere is absorbed by the dust and the combination provides a path for electrical leakage, which hastens the depreciation of the battery. Usually batteries are set away and are entirely forgotten, with the result that a thick layer of dust often collects on their surface.

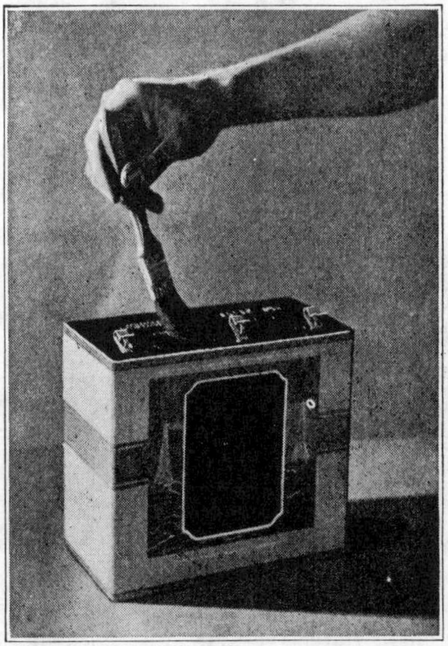

Fig. 107. Considerable Dust Collects on Batteries and This Should Be Brushed off Occasionally

Storage B-Batteries.—Storage B-batteries can also be obtained for providing plate current. They are usually put up in blocks consisting of 24 cells each. Every cell delivers a potential of 2 volts, and the whole block therefore delivers a total potential of 48 volts. Each cell consists of a small glass jar containing an insulator and two plates, the terminals of which project up through the top, which is usually sealed. The electrolyte is usually a sulphuric-acid solution. A small vent hole is provided in the top of each cell to permit the escape of gas formed by charging and to permit the insertion of the nozzle of a hydrometer or syringe, for testing and for adding water.

Most of the objectionable features of the ordinary storage battery are also encountered in storage B-batteries, as, for instance, the electrolytic troubles, plate troubles, and the constant bother of charging them. Besides, they are more or less bulky and may be noisy. Their cost, too, is much more than that of dry-cell B-batteries, and this also accounts for their lack of popularity.

In addition to the above-mentioned troubles, storage B-batteries often become unusable owing to one or more cells becoming defective. In such cases the defective cell can be shunted out by connecting together the two cells on either side of the defective one, as shown in Fig. 108. This method, of course, decreases

Fig. 108. Shunting out a Defective Cell in a Block of Storage B-Batteries

the total output of the block by two volts. Such an emergency method of repair will be satisfactory until the defective cell is replaced with a new one. A defective cell can usually be located by using a voltmeter; every good cell will cause the needle to deflect, but a "dead" cell will, of course, not cause a deflection. It is impossible to use an ordinary charger for charging storage B-batteries, as they require a much lower charging rate and a higher voltage. Many chargers on the market have special taps and a resistance device, which makes them adaptable for charging storage B-batteries. Usually the blocks of cells have to be disconnected and each block must be charged separately. This practice also entails considerable inconvenience, which is not the case when using dry-cell B-batteries. It must be remembered when charging storage B-batteries, that the charging potential must be slightly above the potential of the battery and the current passed should not exceed ¼ ampere.

C-Batteries

C-batteries are used in receivers to provide a voltage on the grid of some of the tubes, to hold them either negative or positive. This grid bias greatly influences the flow of plate current passing between the filament and the plate. Generally 4½-volt C-batteries are used for providing a grid bias on the standard UX 201A or CX 301A tubes, but a UX 171 or CX 371 tube requires a grid potential of 40.5 volts, when the plate potential is 180 volts. It is possible to tap B-eliminator devices to obtain grid voltages, but an exceptionally steady voltage is required for the grid, and this cannot always be obtained from B-eliminators. For steady, dependable service C-batteries are therefore preferable, and under average conditions they will last from four months to a year.

It should be kept in mind that it is better to use too low a value of C-voltage than too high a value, for a high C-voltage

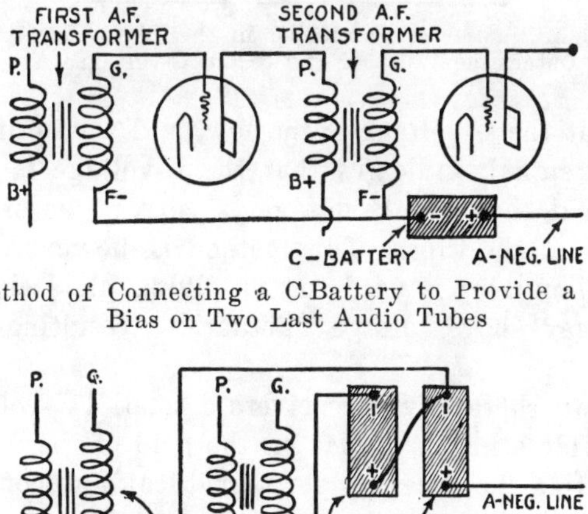

Fig. 109. Method of Connecting a C-Battery to Provide a Negative Grid Bias on Two Last Audio Tubes

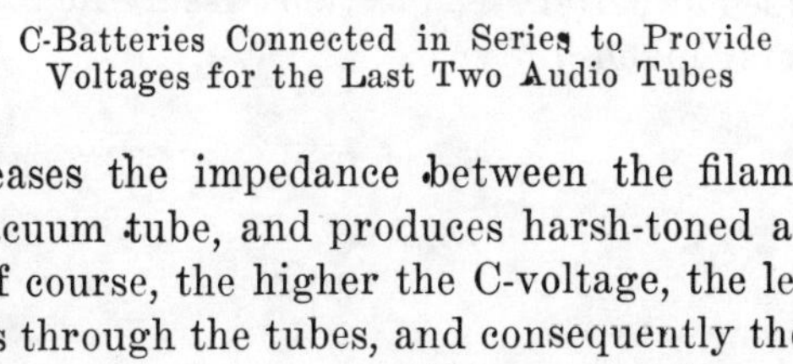

Fig. 110. Two C-Batteries Connected in Series to Provide Different Grid Voltages for the Last Two Audio Tubes

greatly increases the impedance between the filament and the plate of a vacuum tube, and produces harsh-toned and distorted reception. Of course, the higher the C-voltage, the less plate current will pass through the tubes, and consequently the B-batteries will last proportionately longer. When connecting a C-battery to the audio tubes of an old receiver care should be taken to get the correct value as given on the carton in which the tube is sold. Also see Fig. 153, where this information is given. The proper method of connecting a C-battery to two audio tubes of the UX 201A or CX 301A type, using 90 volts of plate potential, is shown in Fig. 109. A higher C-voltage is required for the last audio tube in case a power tube is used, and two separate C-batteries should then be provided, as shown in Fig. 110.

If it is desired to determine whether or not the correct C-voltages are applied in accordance with proportional plate voltages, the milliammeter method should be used, as shown in Fig. 111. A drop of the needle, when reproducing a strong signal,

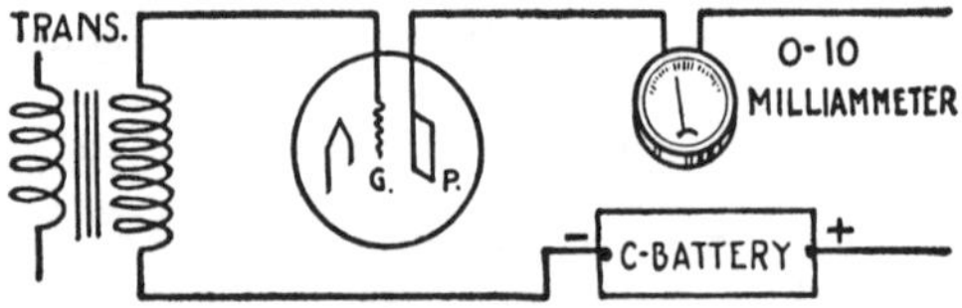

Fig. 111. 0-10 Milliammeter Connected in the Plate Circuit of a Vacuum Tube to Determine Whether the B and C-voltages Are Correct

indicates that the C-voltage is too low, and a rise of the needle, on a strong signal, indicates that the C-voltage is too high, it being assumed that the B-voltage is correct, according to the requirements of the tubes. This method is, however, not reliable in case an eliminator is used for supplying plate current, as an eliminator may show voltage fluctuations resulting from other causes.

In some superheterodyne receivers a small 1½-volt dry cell is used to provide a negative bias on the grid of the detector tube, thus eliminating the necessity for a grid leak and condenser. The positive side of the dry cell is connected to the negative filament line, and the negative side is connected to the center tap of the loop. Reversed C-battery connections result in fuzzy, distorted signals or total inaudibility.

CHAPTER V

BATTERY ELIMINATORS AND THEIR TROUBLES

At the present stage of radio development the average receiver is provided with five, six, or seven tubes in order to obtain maximum tone quality, volume, and range, instead of two, three, or four, as was the case only a few years ago. Along with this increase in the number of tubes used in radio sets has come the need for more power and longer life in batteries. While the battery manufacturers have, to a large extent, kept pace with the ever-increasing demands, producing larger and more durable batteries, after all, this has not solved the problem. The time has come when batteries are being eliminated and radio receivers, except portable sets and those used on the farm, will

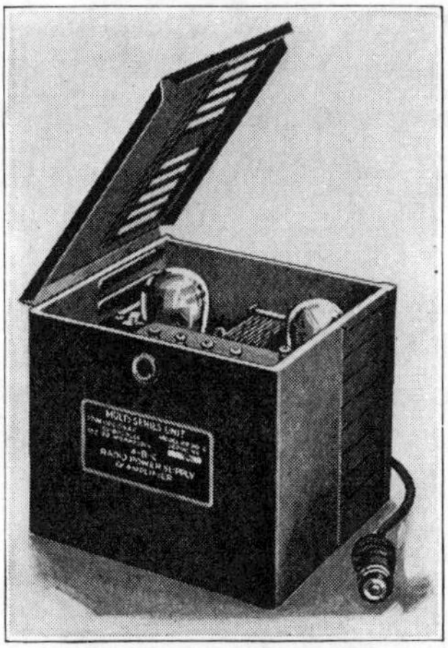

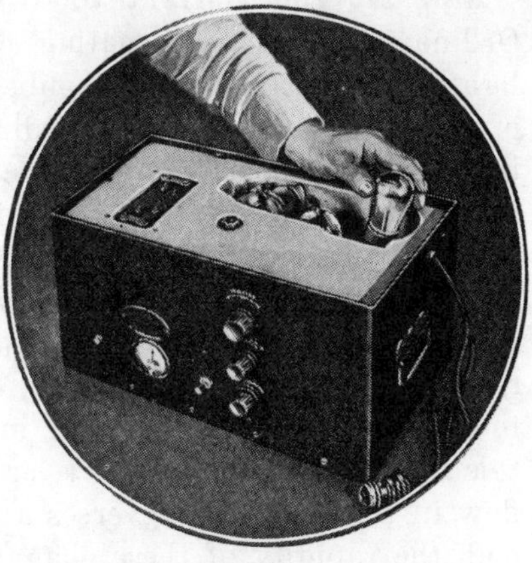

Fig. 112. Two Typical Power Units Operating on 110-Volt Alternating Current, Which Furnish A, B, and C Current for Radio Receivers
(Courtesy of "Popular Mechanics Magazine")

depend entirely on the house-lighting current for their source of electrical supply. With this radical change from batteries to power devices for supplying current, it is necessary for the average owner to acquaint himself with the new lines of progress. Every radio fan should know the general principles involved to some extent, and be able to find and remedy minor troubles which may develop, just as the average auto owner should know something about his car and its operation. In this section we shall not deal with the various troubles encountered with pseudo-eliminators, which have a trickle charger and storage battery in addition to a B-eliminator, and therefore share most of the common charger and battery troubles, but we shall go into trouble shooting in respect to real eliminators, which change the house-lighting current into the steady, direct current required by a radio receiver. The theory of eliminators will be gone into only so far as is really necessary for the fan to understand the function of the parts so that he will be better able to find any trouble which may exist from time to time. Two eliminators providing A, B, and C-supply are shown in Fig. 112.

Construction and Operation of Battery Eliminators

Raw 110-Volt Current Unsatisfactory for Tube Operation.— Ordinary 110-volt alternating current (a.c.) from the house-lighting circuit cannot be applied directly to the vacuum tubes of a radio receiver, for the ordinary type of radio vacuum tube is designed to operate on a direct current of 3 to 6 volts for the filament, and a direct current of 18 to 135 volts for the plate. However, raw a.c. can be used satisfactorily on most power amplifier tubes. It is also impractical to use a.c. for the plate voltage of the tubes, even if suitable resistances are used to reduce it to the proper voltage, because the a.c. would give rise to constant, undesirable hum in the receiver. The reason for this hum is that the polarity, or rather the direction of current flow in the a.c. lines, reverses a given number of times per second, the rapidity of these changes or reversals of current being designated by "cycles." A 60-cycle current surges forward and backward alternately 120 times per second, one surge and its accompanying reversal being called a cycle. Such a frequency, when applied to a radio tube, would be amplified considerably, and as this frequency is within range of the human ear, being called audio frequency, it would cause a constant hum in the receiver, making reception very unsatisfactory if not entirely impossible. Commutator and slip-ring noises also interfere with reception.

Unfiltered Rectified A.C. Also Unsatisfactory.—Thus the a.c. must first be converted to direct current (d.c.), and furthermore, it must first be stepped up to a higher voltage than can be obtained directly from the house-lighting circuit, which is done by means of a transformer. Still, the conversion of a.c. to d.c., or rectification as it is called, would only result in a continuous rise and fall of the d.c., the rapidity of these fluctuations corresponding to each cycle of the a.c., and the hum would therefore still persist. So means must be employed to smooth out these

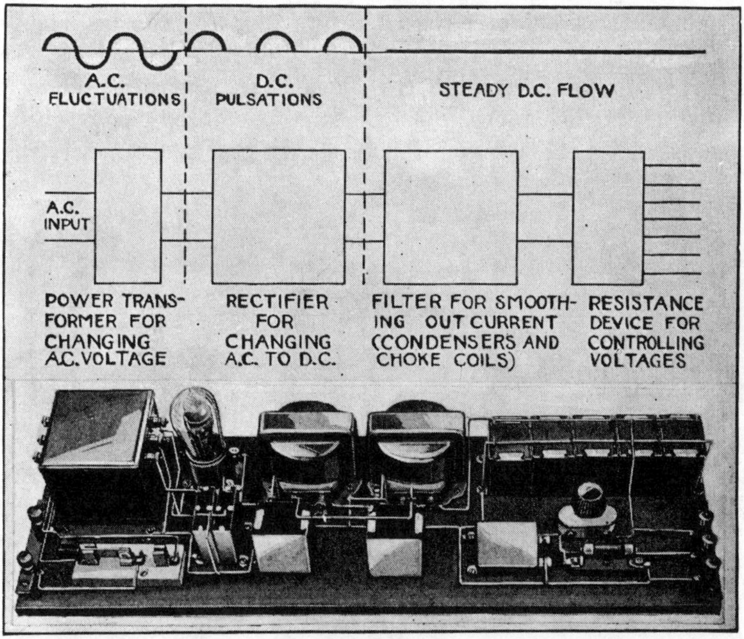

Fig. 113. Instruments Used in a Battery Eliminator Arranged in Order of Their Operating Sequence. The Diagram Shows the Four Sections of an Eliminator and How It Changes Alternating Current to Continuous Direct Current

fluctuations. Besides this, an A, B, and C-eliminator must have a number of voltage taps of different values; first, for the plate circuits of the radio-frequency tubes, the detector, the audio-frequency tubes, and perhaps a power tube; second, for the filament circuit to light the tubes, and also for the grid-biasing circuit.

Accordingly, there are four parts in every eliminator, namely, a step-up transformer, a rectifying device, a filtering device for eliminating fluctuations, and resistance devices across the output. The mutual arrangement of these parts and their individual functions are shown in Fig. 113. The transformer is designed to deliver a secondary voltage far in excess to that actually required at the B-output terminals of the eliminator, in order to compensate for the energy losses and the consequent drop in voltage in the rectifier and filter systems.

Rectifying Devices.—The rectifying device is connected to the output of the transformer. There are four common methods of accomplishing rectification in battery eliminators, namely, by means of metallic rectifiers, by chemical rectifiers, by filament-heated rectifying tubes, or by means of ionized-gas rectifying tubes, each of these devices being shown in Fig. 114. The chief

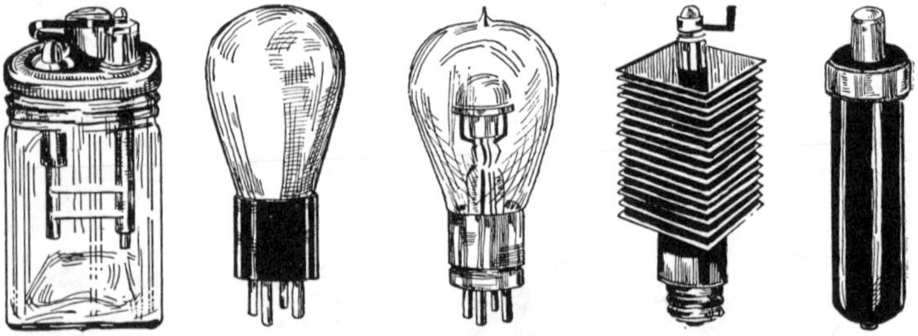

Fig. 114. Various Types of Rectifier Units Used in Battery Eliminators and Chargers for Changing Alternating Current to Direct Current

objection to the use of chemical rectifiers is that a liquid is used, and this is naturally subject to evaporation, and must be replaced now and then. Also, the electrodes used in chemical rectifiers are subject to decomposition and must be renewed occasionally. Sometimes corrosive vapors are given off, which is another objectionable feature. The filament-heated or electron tubes have the disadvantage of being rather short-lived, and from this standpoint they are more or less expensive. However, they have a capacity for delivering a considerable amount of rectified current and are dependable. The ionized-gas rectifiers are very popular at the present time, particularly for B-eliminators, and therefore they are used in the great majority of eliminators. Two well-known rectifiers of this type are the Raytheon tube and the Q. R. S. tube, the latest type used for B-eliminators having only a capacity of 85 milliamperes, at 200 volts continuous load, while the safe a.c. load which can be applied to each anode is set at 350 volts. This type of rectifying tube has no filament to burn out, and its life is therefore much longer than that of the filament-lighted rectifiers. There is also a 350-milliampere Raytheon tube and a 400-milliampere Q. R. S. tube designed to supply current for .25-ampere, 5-volt tubes.

The Filter System.—After the current has been rectified, its fluctuations must be smoothed out by means of a filter system. Although there are different methods of filtering, the generally used filter circuit in radio eliminators consists of two choke coils and a number of condensers, examples of which are shown in details 1 and 2 of Fig. 115. Choke coils are simply cores of laminated-iron strips, on which a great number of turns of wire are wound, the size of the core, the size of the wire, and the number of turns of wire determining the output capacity and the efficiency of the eliminator. If a small core and a large number of turns of fine wire are used to smooth out effectively the pulsations of the current, then the current output is greatly decreased, and on the other hand, if a large core, large-size wire, and fewer turns are used, the choke will not be so effective in eliminating the pulsations, although the amount of current passed will be greater.

The condensers, which are shunted around the choke coils, serve as "storage tanks" for the excess current not used by the receiver, and to smooth out the remaining fluctuations of the d.c. This stored electricity is held in readiness to be used in case of an instantaneous overload on the circuit, as, for instance, when an especially loud note is reproduced, which often consumes much more current than the choke coils can pass, and this extra energy is then supplied by the condensers. Such a condenser discharge does not last long, usually not more than two seconds, after which the condenser is "empty," but it is "refilled" again when the receiver has resumed its normal operation. The output current of the eliminator is, after passing the filter system, an even, undisturbed direct current.

The Resistors.—The positive output line is tapped at various points, and the current obtained at each tap is passed through a resistance unit, either variable or fixed, which decreases the voltage to the proper value for the plates of the detector, radio-frequency, and audio-frequency tubes. Both A and B power supply is similarly obtained through such resistors, an example of which is shown in detail 3 of Fig. 115.

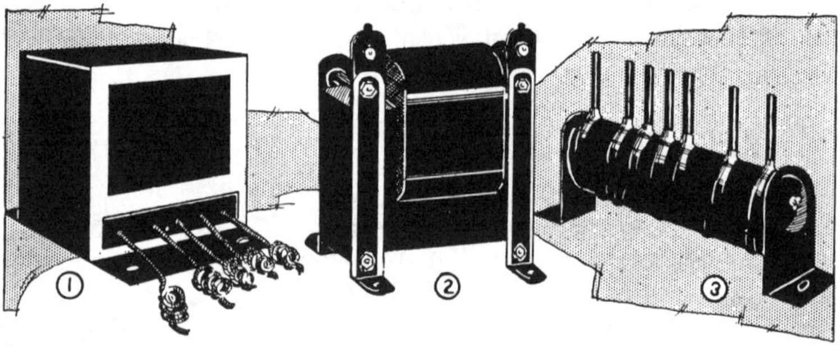

Fig. 115. Most Essential Instruments Used in Battery Eliminators: (1) Large Block Condenser Used in Filter Section. (2) Choke Coil for Smoothing out D. C. Pulsations. (3) Unit Resistor Strip for Obtaining Different Voltage Values

Using A-Eliminators

In general, the principle of an A-eliminator is the same as that of a B-eliminator, except that the output voltage of an A-eliminator must be lower than that of a B-eliminator. In the design and construction of A-eliminators the greatest problem is encountered in the filtering device, as it must be able to pass considerable current without overheating.

Series Connection of Tubes for A-Supply.—When filament current is obtained from an eliminator it is usually necessary to connect the tubes in series, especially in case 5-volt tubes are used. In making the wiring changes in the filament circuit, the last audio tube should receive the positive output of the eliminator supply, while the detector should be connected to the negative end. This is because the hum characteristic of power devices is usually more pronounced at the positive end of the eliminator output than at the negative end, and the hum is least noticeable in the receiver if the terminals are connected up as just explained. The voltage required for the operation of any number of tubes connected in series is equal to the product of the number of tubes and the voltage on which they operate separately. Thus six 3-volt tubes require 18 volts, and six 5-volt tubes require 30 volts. In most eliminators the output voltage can be varied by means of suitable resistors, instructions for the use of such resistors being usually supplied by the manufacturer of an eliminator.

Grid-Biasing Voltages Direct from Filament Line.—The correct grid bias for each tube can be obtained direct from the filament line, except in case a bias of a higher value than the maximum filament voltage is required. The voltage of the line is decreased 5

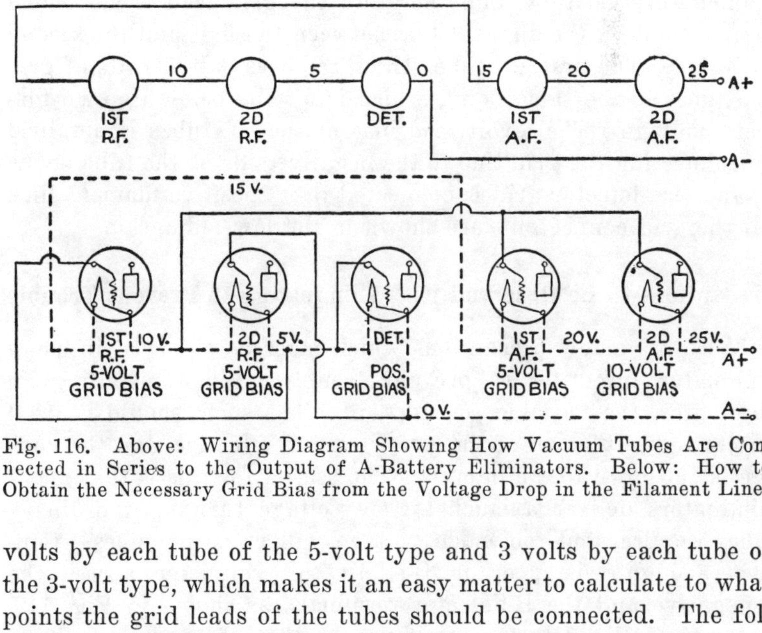

Fig. 116. Above: Wiring Diagram Showing How Vacuum Tubes Are Connected in Series to the Output of A-Battery Eliminators. Below: How to Obtain the Necessary Grid Bias from the Voltage Drop in the Filament Lines

volts by each tube of the 5-volt type and 3 volts by each tube of the 3-volt type, which makes it an easy matter to calculate to what points the grid leads of the tubes should be connected. The following rule should be observed: The grid lead of a tube is connected to a point on the filament line where the voltage is lower than the negative side of the filament of the tube in question, the difference in voltage between this point in the line and the negative side of the tube whose grid is thereby biased being equal to the amount of bias desired. This rule will be better understood by carefully studying Fig. 116. Here is shown a 5-tube receiver equipped with 5-volt tubes. It is assumed that the following grid-biasing voltages are necessary: 10 volts for the second a.f. tube, 5 volts for the first a.f. tube, 5 volts for each of the two r.f. tubes, while the detector grid is positive. As the tubes are connected in series, a maximum filament voltage of 25 volts is necessary to force a .25-ampere current through the five tubes, the positive end being connected to the last a.f. tube and the negative end to the detector, as shown in the upper diagram. There will then be a potential of 20 volts between the first and second a.f. tubes, 15 volts between the first a.f. and the first r.f. tubes, 10 volts between the first and second r.f. tubes, and 5 volts between the second r.f. and detector tubes. The grid bias on the second a.f. tube must be 10 volts below the negative filament of this tube, and therefore the grid lead is connected to any point on the filament line between the first and the second r.f. tubes. The first a.f. tube, which requires only 5 volts of grid bias, goes to the same point, which is 5 volts below its own filament voltage. The 5-volt grid bias of the r.f. tubes is obtained by connecting the grid line to the negative side of the tube below it, and the detector grid is connected to its positive filament line. All the grid connections are shown in the lower diagram.

Precautions to Be Observed with Eliminators to Prevent Trouble

There are a few precautions which must be observed in using eliminators in order to prevent trouble. When installing an eliminator, the variable resistors, if it has such, should be well unscrewed so that all of their resistance is effective, and the voltage is cut down to a minimum. The reason for this is that many eliminators deliver a much higher voltage than most ordinary tubes require, and too much voltage will ruin the tubes. Connect a high-resistance, or B-eliminator voltmeter across the B-negative and the B-detector terminals, as shown in Fig. 117, and adjust the detector resistance so that the correct voltage for the detector tube is obtained. Then proceed in the same way with the intermediate and amplifier taps. Only a high-resistance voltmeter can be used for this purpose satisfactorily, as a low-resistance voltmeter consumes so much current that it materially reduces the voltage while the test is being made, resulting in an erratic reading and a faulty adjustment.

The receiver should always be turned on first; then the eliminator should be switched on. This practice must always be observed except in eliminators using a glow tube to stabilize the output used for the plate circuits of the r.f. tubes. The result of neglecting this precaution may mean injury to the eliminator, as its internal pressure will temporarily be so great, when turned on without a load, that one or more of the condensers may be

Fig. 117. Only a High-Resistance Voltmeter Can Be Used for Accurately Testing the Output Voltage of a Battery Eliminator
(Courtesy of "Popular Mechanics Magazine")

punctured, which means a short circuit and total inoperation of the eliminator. If the receiver has a separate switch, as is usually the case, this should always be left on, for the reason just mentioned. Furthermore, if at any time changes have to be made in the wiring of the receiver, loose connections tightened, or parts removed, one should be absolutely certain that the B-supply is turned off. The most positive way is to disconnect the plug from the light socket. Otherwise serious damage from accidental short circuits may result.

Cause and Cure of Humming

Poor Adjustment.—The most common trouble experienced with eliminators is humming. This may, of course, be due to poor design in the eliminator, but taking it for granted that the eliminator is a good one, there are several other causes which may give rise to this trouble. Often the hum results from faulty adjustment of the variable controls; in fact, few eliminators will work satisfactorily at once without manifesting a noticeable hum, until they are properly adjusted. On most eliminators the adjustment can be made by first turning the variable controls entirely to the left or "off" position and then gradually turning them up toward the right until a good, clear signal is obtained on the receiver. The detector voltage is the most critical, and it should be kept as low as possible, just above the point where signals become "fuzzy." Another reason for keeping the detector voltage as low as possible is to reduce its tendency to be "microphonic," a condition to which most "hard" tubes are susceptible, and which is aggravated by a high plate voltage. Too much plate voltage on the detector tube will also produce harsh-toned reception. The detector-plate voltage is first adjusted and then the amplifier voltages. It must be remembered, however, that the variable resistances exert a mutual effect on each other, and therefore the detector control must be readjusted after the amplifier control has been set correctly. After an eliminator has once been correctly adjusted, do not change it or allow anyone to meddle with it. This is often a great temptation for the radio fan; the more variable controls there are to handle, the more he likes to play with them. It should be borne in mind, however, that constant readjustment not only is a waste of time, but the variable controls are not made for constant rotation. This handling may cause them to become defective, and then no marked improvement can be obtained by manipulating them; signals will become fuzzy and distorted, or no reception will be possible at all.

Reversing Plug, and Grounding Eliminator.—Often the hum can be remedied by merely reversing the plug which connects to the light socket. Grounding the eliminator is also effective, and unless the receiver is especially designed to operate on an eliminator, it will usually be necessary to connect the B-negative terminal of the eliminator to the ground or A-negative terminal of the receiver, provided the A-negative and B-negative lines on the receiver are connected. If the A-positive and B-negative on the receiver are connected, the wiring should be changed so that the A-negative and B-negative are connected. When this precaution is omitted, there is often a pronounced a.c. hum. The metal case in which almost every eliminator is inclosed also serves as a shield, and this should be connected to the ground. But the B-negative terminal and the metal case of the eliminator can be grounded by the same wire.

Inductive Effects of Eliminator.—Proximity of the eliminator to the receiver is also a frequent cause of hum, owing to induction, and this is often the case with console installations, where the eliminator occupies the lower compartment directly under the receiver. Of course, the best method of preventing the hum is to install the eliminator away from the receiver, but this is often undesirable. Thus, in such cases, shielding the receiver must be resorted to. The method of doing this is completely described in the section entitled, "Complete Shielding of the Receiver."

Induction from A.C. Circuit.—Sometimes a pronounced a.c. hum is caused by induction from the house-lighting circuit, which can also be reduced considerably by shielding the receiver, although the easiest remedy is to place the receiver in another location farther away from the offending house wires. Also be sure to remove from proximity to the receiver all portable lamps and other electrical appliances operating on the house-lighting circuit, for proximity of these devices may cause a hum.

Poor Filtration in Eliminator.—If a hum is caused by poor filtration in the eliminator and is carried to the receiver through the leads, it can sometimes be cured by shunting a number of by-pass condensers around the output terminals of the eliminator. It may be necessary to experiment a little in order to get condensers whose capacities are most effective for reducing the hum. Condensers ranging from 1 to 4 mfd. in capacity can be used for this purpose, the smallest one being connected from the B-negative to the B-detector terminals, the second from the B-negative to the B-intermediate terminals, and the third from the B-negative to the B-amplifier terminals. After the correct values have been found, the condensers should be permanently wired to these terminals. In most cases this method will be found quite effective.

Defective Buffer Condenser.—A hum, which becomes noticeable only after the eliminator has been used for some time, and which did not exist at first, may be caused by a faulty buffer condenser, the position of which is indicated in Fig. 113. This trouble can be cured only by the replacement of the offending condenser. The method of testing it will be explained in a later paragraph. A defective rectifying tube sometimes causes a hum, and should be replaced with a new one.

Mechanical Hum or Vibration of Eliminator.—Eliminators frequently produce more or less annoying mechanical vibration or hum, which cannot be heard in the loudspeaker, but will, nevertheless, interfere with reception to some extent. This is sometimes due to a loose core in the transformer, and if this is of the open type, it can usually be tightened with very little trouble. However, many transformers are of the enclosed type and are set in a metal shell, into which an insulating compound is poured to completely cover the windings. In such cases it may be necessary to send the eliminator back to the manufacturer for repair. If this is not desired, install the eliminator in the basement, and set it in a position directly under the receiver so that the leads can be brought through a small hole drilled in the floor near the baseboard. Sometimes the metal case of an eliminator, which is made of sheet metal, is not assembled securely, and the transformer vibration causes the case to vibrate quite audibly. The remedy consists in tightening the case. Often eliminator hum of a mechanical nature can be prevented by setting the device on a sponge-rubber cushion.

Crackling Noises

Short-Circuited Buffer Condenser.—Crackling noises which are more or less intermittent and give a series of well-defined clicks can often be traced to loose connections or broken leads between the eliminator and the receiver, but may also exist in the eliminator itself. Sometimes the cause of crackling noises is an open or short-circuited buffer condenser, which is shunted across the anodes of an ionized-gas rectifying tube to prevent arcing. This defect may permit arcing inside of the tube with consequent crackling noises in the receiver, which somewhat resemble static disturbances. The remedy is a new condenser.

Arcing Between Anode Leads of Tube.—Arcing between the two anode leads outside of the cup in a gas-filled rectifier tube may also cause this trouble, even though the buffer condensers are still in good condition. This trouble is caused by a break in the lava insulating tubes, and in this case the substitution of a new tube is the only remedy. Another cause of crackling attributable to the eliminator is arcing between the prongs of the rectifier-tube socket, and, in very rare cases, across the output leads of the secondary winding of the transformer. The prongs on the socket come very close together, and if they are slightly loose, or if there is a path across them such as provided by a daub of soldering paste or the like, which may start arcing, considerable damage may result. Therefore, before using an eliminator, the socket prongs should be carefully examined, and all traces of flux or moisture should be wiped off before inserting the rectifier tube. A fourth cause of crackling noises can be traced to defective resistors. Many resistors are unable to stand up under the high voltages provided by eliminators, and at the same time to pass enough current without overheating. However, more sturdy resistors for this purpose are being put on the market, and within a short time this trouble from resistors will be less frequent than it is at present.

"Motorboating" and How to Cure It

What "Motorboating" Is.—"Motorboating" is a peculiar puttering sound often noticed in receivers operated on battery eliminators, and sometimes even in receivers operated on batteries. The noise somewhat resembles the steady "putt-putt" of a motorboat engine, and its frequency may be anywhere from 2 to 30 times per second. At higher frequencies than this the noise ceases to be motorboating and becomes a low howl, although in both cases the cause may be the same. Another effect due to the same cause is the complete blocking of the signal. In motorboating the alternating increase and decrease in the intensity of the plate current, especially that of the last audio tube, may be so great that a flickering of the filament of this tube, in synchronism with the noise, may be observed. Because of the inherent tendency of resistance-coupled receivers (those having resistors between the audio tubes instead of transformers) to motorboat, it is erroneously held by many that these receivers cannot be operated satisfactorily on battery eliminators. Transformer-coupled receivers are less subject to this trouble but in some cases they will motorboat, even when operated on ordinary batteries, which proves that the trouble is not always and entirely due to eliminators. Impedance-coupled receivers (those that have impedance coils between the audio tubes instead of transformers or resistors) are least of all subject to this trouble.

Motorboating should not be confused with an a.c. hum, which is noticeable in a greater or less degree in nearly every receiver operated from an a.c. power device. The two noises have different causes and can readily be distinguished; the a.c. hum always occurs at a constant frequency in synchronism with the number of cycles of the lighting current, while motorboating may be evident at various frequencies. Besides this, the a.c. hum is usually quite persistent and regular, while motorboating is more or less irregular and periodical, depending on various conditions. Some of the causes of motorboating lie in the eliminator, while others can be found in the receiver. It must be remembered that all receivers will not operate equally well on battery eliminators, and in some cases it is necessary to change the receiver to meet the conditions arising from the use of an eliminator. At other times the eliminator is poorly designed or defective, and may be the sole cause of the trouble. Seldom, however, will a receiver function properly on an eliminator until the latter has been correctly adjusted. It is generally conceded that the most common cause of motorboating is the high internal impedance of the filter section of the eliminator. Faulty resistors and slow-leaking condensers, not able to retain a charge, also cause motorboating. If the trouble has been evident only for a short time, a defective resistor may be discovered, especially if it has been handled considerably. Variation in the potential of the house-lighting current, which is especially prone to occur on small distribution systems, causes a corresponding fluctuation of the voltage in the eliminator, and such a surge may set up motorboating. Sometimes the failure of the rectifying device to operate properly, or the lack of stability of the rectifying device will cause the trouble.

It has been found that motorboating can often be remedied by shunting resistors and condensers across the output terminals of the eliminator, as shown in Fig. 118. In some cases it will be found helpful to shunt a glow lamp, such as a UX 874 or CX 374 across the B-negative and the B-90 or intermediate terminals, which supply the radio-frequency tubes; see Fig. 131. This controls the voltage applied to the radio-frequency tubes and permits them to operate uniformly at all times. Proper control and adjustment of their plate voltage, and that of the detector, is necessary, as excessive plate voltage on these tubes may cause motorboating. The voltages applied by the eliminator should therefore be checked up occasionally, and it is advisable for the fan to provide himself with a high-resistance voltmeter of the type specially used for this purpose, as an ordinary voltmeter will not give correct output readings of eliminator voltages. With the aid of a high-resistance voltmeter, however, the fan can readily adjust the output voltages of his eliminator correctly according to the exact requirements of the various tubes.

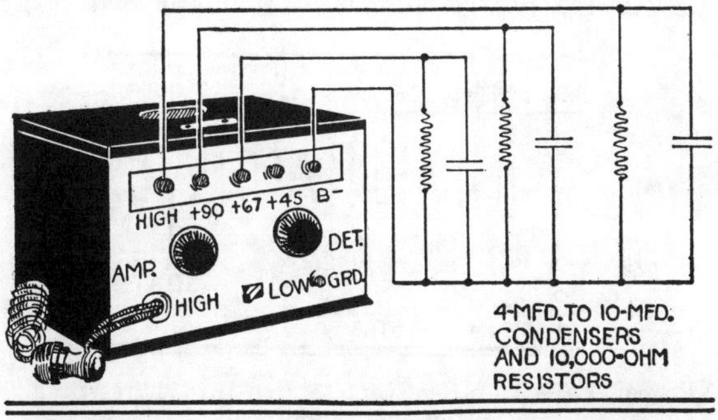

Fig. 118. "Motorboating" Caused by B-Eliminator Can often Be Reduced by Means of a Number of 4 to 10-mfd. Condensers and 10,000-Ohm Resistors

Inserting Choke Coil in Receiver.—An effective remedy for motorboating consists in connecting a radio-frequency choke coil in series with the detector plate and the first resistor, in case of a resistance-coupled receiver, and between the detector plate and the P-terminal of the transformer, in case of a transformer-coupled receiver, as shown in Fig. 119. Substituting a 1 or 2-megohm resistor for the usual .1-megohm (meg.) resistor in

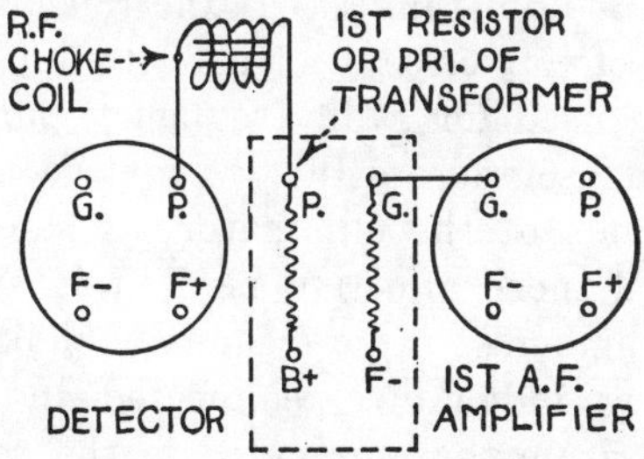

Fig. 119. Method of Eliminating Motorboating by Connecting an R.F. Choke Coil in the Plate Circuit of the First Detector

the plate circuit of the detector, in case of a resistance-coupled receiver, will also help materially to reduce the trouble. The values of the resistors in the various grid and plate circuits of a resistance-coupled circuit using high-MU tubes, which have been found satisfactory to prevent motorboating, according to a well-known manufacturer, are shown in Fig. 120. In case these changes show slight signal-interference effects, when using high-MU tubes, remove the tube in the first stage and substitute a regular UX 201A or CX 301A, or similar type of tube. If a UX 112 or CX 112 power tube is used in the last stage of audio,

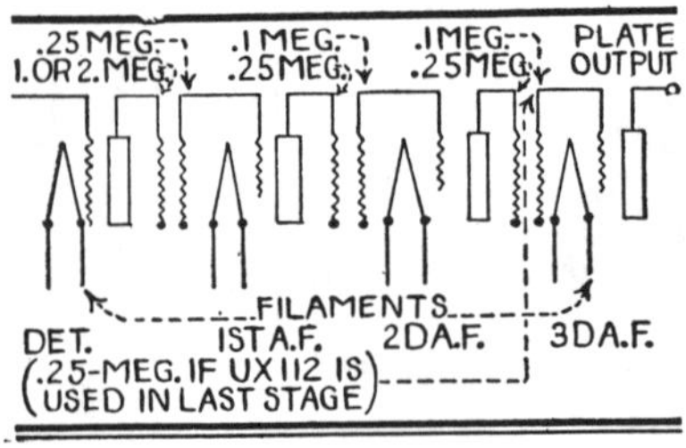

Fig. 120. Proper Values of Resistors to Use in a Three-Stage, Resistance-Coupled Amplifier

the resistor in its grid circuit should be of .25-meg. value. In resistance-coupled receivers not having a three-stage amplifier unit of the Daven or similar type, it will be found that the substitution of a well-designed transformer or impedance unit will usually be quite effective in preventing the trouble. When making this change, however, care must be taken to wire the transformer or impedance unit up correctly, according to the markings on the terminals.

Chokes in the B-supply Lines.—If the remedies just mentioned are not entirely effective, connect a 30 to 100-henry choke coil, which is the size used for radio-frequency purposes, and also a 4 to 8-mfd. condenser across the B-negative and the maximum B-positive terminals of the eliminator, as shown in detail 1 of Fig. 121. A still more effective method is shown in detail 2, which shows a number of radio-frequency choke coils of approximately 100-henry inductance, connected in series with each plate-supply line from the eliminator to the receiver.

Separate B-Battery for Detector Voltage.—In case of a superheterodyne or other large receiver, it will be found advantageous to make use of a separate B-battery to supply detector-plate voltage. Of course, this plan partly defeats the purpose of B-eliminators, but it may nevertheless be the best remedy. It must be taken into consideration that one remedy may work better than another, depending on the type of receiver, and the best cure for motorboating in any particular set can best be found by trying out the various methods outlined in this section, one by one, until the trouble disappears.

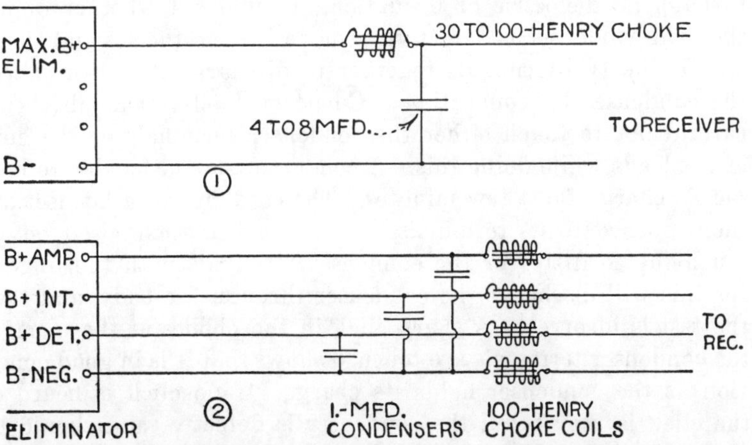

Fig. 121. Two Methods of Reducing Motorboating Caused by Eliminator. (1) Use of a 30 to 100-Henry Choke Coil and a 4 to 8-mfd. Condenser. (2) Choke Coils in Series with All Output Leads

Incorrect Grid Bias.—If no improvement is noticed after trying all the methods already mentioned, then the trouble may be found in the receiver itself, which should be gone over thoroughly. Incorrect grid bias, or voltage on the grids of the various tubes, is a condition in the receiver that is especially conducive to motorboating. Be sure to have the correct C-negative voltage on the grid of the power tube. This voltage is given on the sheet of instructions accompanying the tube when it is purchased, or is printed on the carton in which the tube is sold. In general, it must be borne in mind that radio-frequency and audio-frequency tubes should have a negative bias, while the detector usually has a positive bias. It is a good idea to connect a variable resistor, such as a Clarostat, in the grid circuits, so that the voltage can be correctly regulated. Too much grid voltage is likely to set up motorboating, and if a variable resistor is inserted in the line such excessive voltage can readily be reduced. The proper method of connecting a C-battery in the grid circuit of a power tube is shown in Fig. 110. Be sure that the negative terminal of the C-battery connects to the grid terminal of the socket of the power tube.

Defective Grid Condenser.—Another cause of motorboating is a defective grid condenser, which slowly dissipates energy through its dielectric or insulation. To find out whether or not the grid condenser is defective, remove it from the set and short-circuit the two terminals together to discharge it. Then charge the condenser by connecting a C-battery to its terminals, being careful not to touch either the condenser terminals or the ends of the leads while doing this. A condenser should be able to hold such a charge for a few minutes. The cord tips of a headset are then touched to its terminals, care again being exercised not to touch the cord tips or the condenser terminals with the fingers, for this will discharge the condenser through the body, and then the test is incorrect. A sharp click in the phones at the moment the condenser terminals are touched shows that it is in good condition as the condenser holds its charge. If no click is heard or the click is very faint, the condenser is defective and should be replaced with a new one. Humidity conditions have considerable effect on open condensers, and in moist places the dielectric may absorb enough moisture to cause it to become leaky. The enclosed type of fixed condenser, such as is enclosed in a bakelite shell, is impervious to moisture. It is a good plan to substitute condensers of this kind for the open type, being sure that the new ones are of the same values as the discarded ones. Too high a value of grid resistance may contribute to the tendency to motorboat, and if this is the case, a resistor of lower value should be substituted.

Reversed Primary Winding of A. F. Transformer.—Motorboating is sometimes caused by a reversed condition of the primary winding of an audio-frequency transformer. This reversed condition may be attributed to incorrect internal connections between the winding and the terminals on the transformer shell. All that is necessary, in order to remedy this trouble, is to reverse the transformer connections on the outside, that is, the B-positive line will then connect to the P-terminal of the transformer, and the lead from the P-terminal on the preceding tube socket will connect to the B+ terminal on the transformer.

Unmatched Impedance of Speaker and Power-Tube Plate Circuit.—Motorboating can also be caused by a lack of matched impedance of the magnet winding of the loudspeaker and the plate of the last tube. Many loudspeaker manufacturers are trying to match the impedance of their speakers to that of the generally used power tubes, but power tubes have different characteristics and new ones are constantly coming out on the market, which makes it impossible to obtain perfect matching in this way. Besides, many fans have loudspeakers before they install power tubes in their sets, and it is therefore necessary for them to use other methods of overcoming motorboating caused in this way. There are several types of output transformers and output impedances which can be added to any receiver, as fully explained in the section on Reproducers. In output transformers the plate output of the power tube is passed through the primary winding, and the loudspeaker terminals are connected to the secondary windings of the output device. An output impedance has the advantage of preventing possible coupling between the plate circuit of the last tube and that of the other tubes of the receiver through the medium of the output resistance devices of the eliminator.

Total Inoperation of B-Eliminator

In case no signals can be obtained from a receiver and there is no doubt that the receiver itself is in good condition, the trouble may be found in the eliminator. Fig. 122 shows the parts where trouble is likely to occur. First see that the leads from the eliminator to the receiver are unbroken and are securely connected.

Faulty Rectifier.—Examine the rectifier, for this may have become defective; if it is of the electrolytic type, the electrolyte may have evaporated, or it may be too low for the proper operation of the device. The electrodes have perhaps become decomposed, which necessitates renewal. If the eliminator uses a filament-lighted tube rectifier, see whether or not it is lighting properly. It must be remembered that these tubes have a definite life and must be renewed sooner or later. Even the ionized-gas rectifying tubes, such as the Raytheon or the Q. R. S. tubes, which have an exceedingly long life, there being no filament to burn out, become exhausted eventually. Normally, they give off a bluish flash at the edge of the "cup" at the moment that the eliminator is turned on, and this flash can usually be noticed. Sometimes the tube may still be defective, even though it shows signs of life. It is perhaps the best practice for the fan to provide himself with two tubes, so that if he suspects that the one in use is defective, he can substitute the other one.

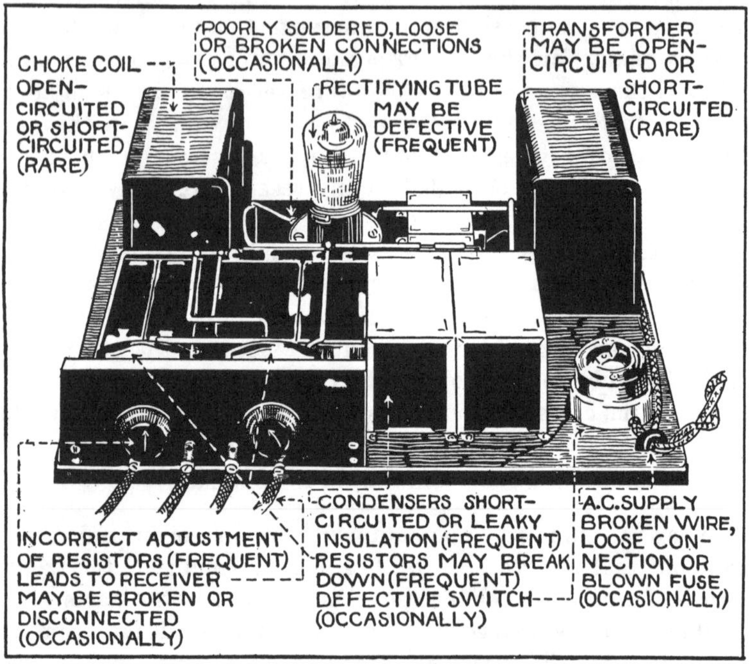

Fig. 122. A Typical B-Eliminator Showing All the Instruments Which Comprise It. The Arrows Point to Some of the Places Where Troubles Are Apt to Develop

Purchasing a New Rectifier.—When purchasing a new rectifier, be sure to get one of the same type as the defective one. If you get one of a higher or lower capacity, the eliminator will not work properly, because the other parts are designed just for the amount of current produced by the original rectifier. For instance, many eliminators, especially the old models, were designed to operate on a type B Raytheon tube, which has an output of 60 milliamperes. When replacing one, some dealers will try to persuade the fan to use a type BH tube of the same make, telling him that it has a greater capacity than the old type and is more durable, etc. The latter type, it is true, has a greater capacity, but if the eliminator is not designed for it, it will not give satisfactory results.

Transformer Trouble.—After the rectifier has been carefully examined, listen for a slight transformer hum. This will tell whether or not the transformer is working. If no hum is noticeable, it is probable that the transformer winding is open-circuited or the transformer is not receiving current from the a.c. circuit. However, open-circuited transformers in eliminators are quite rare, as they are usually designed to stand up under heavy loads. To be certain that a transformer is defective or not, make a test for an open circuit. Some types of transformers used in eliminators have a separate winding to provide current for lighting the power tube of the receiver. In case the power tube does not light, and one is certain that the filament is not burned out, and also certain that the transformer is getting current from the lighting circuit, then the winding, which supplies the filament current for the power tube, may be open-circuited.

Lack of A.C. Supply.—Lack of a.c. supply may often be the cause of trouble if a slight transformer hum is not noticeable, and no output voltage can be obtained from the eliminator. Test the light socket to determine whether or not the supply is being delivered to this point, for it is possible that a fuse of the lighting circuit has been burned out. If the fuse is in good condition, the light socket itself may be defective, even if it makes good contact with the plug occasionally. Sometimes the spring contact in the light socket does not make a good electrical connection with the end of the plug. In this case, turn off the current and pull the prong out so that it will make a better contact. If light can be obtained from the socket, there is perhaps a break in the cord connected to the eliminator, and such a break can usually be found at the plug end. If the cord is broken, it is an easy matter to repair it. Cut the wire at the break, bare both ends for about 2 in., and splice them together, after which they are soldered to make them electrically and mechanically secure. Friction tape is then wound over the splice to insulate and protect it.

Faulty Choke Coils.—If a transformer hum is noticeable, but there is no output voltage across two or more of the eliminator terminals, which can be determined by the needle deflection of any 0 to 150-reading voltmeter held across the B-negative and the B-detector, B-intermediate and B-amplifier terminals successively, then the trouble may be in a broken-down or open-circuited choke coil, although this is rarely the case. To find out which choke coil is at fault, get a pair of headphones and hook up a 22½-volt B-battery in series with the headphones, as shown in Fig. 123. Before applying the free ends of the leads to the coils, remove the eliminator plug from the light socket. Disconnect the choke coils if there is any possibility of a shunt circuit around them, for the coils should be tested separately and there should be no path for current over a shunt. After the coil to be tested has been disconnected, it is short-circuited to free it from any residual charge which it may hold. The free ends of the tester are touched to the ends of the coil, and a click should be heard in the phones. The absence of a click indicates an open circuit, and the coil must then be replaced with a new one of exactly the same design.

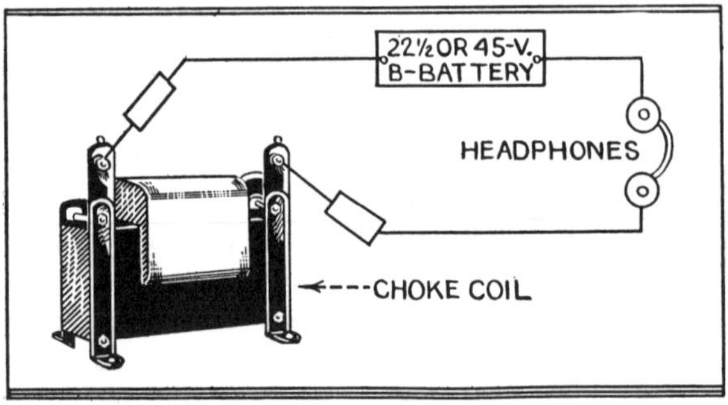

Fig. 123. Method of Using Headphones and a B-Battery to Make a Continuity Test on a Choke Coil Used in Eliminators

Defective Condensers.—A punctured condenser may also prevent the eliminator from functioning. Broken-down condensers, which permit an unrestricted path for the current to flow directly through them, frequently cause eliminator troubles. To test a condenser, it is first removed from the eliminator, as shown in Fig. 124. It is then discharged by holding a wire or a piece of metal across its terminals, for the charge it contains may cause a click in the phones, which will give a false impression that the condenser is short-circuited. If the condenser is in good condition, there will be little or no click the second time that its terminals are touched, but if it is short-circuited, a distinct click will be heard in the headphones each time the condenser terminals are touched, and again when the tips are removed. When making two or more tests on one condenser, be sure to hold the testing tips to the same terminals of the condenser each time. The absence of a click does not always prove that the condenser is in good condition, for there may be a slow leakage of charge through its insulation. To find such a leak remove the B-battery from the headphones and charge the condenser by holding the bared ends of the wires connected to a battery, to the terminals of the condenser for a few seconds. Be careful not to touch either the condenser terminals or the bare ends of the wires, for this will dissipate the charge. After the condenser has been charged in this way, wait for a few moments and then touch the same tips

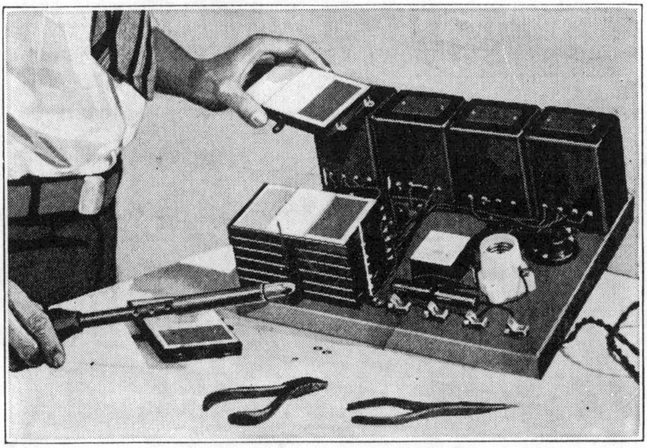

Fig. 124. Removing a Condenser from an Eliminator Prior to Testing It with the Headphone Tester

(Courtesy of "Popular Mechanics Magazine")

of the headphones to the terminals of the condenser. If only a faint click is heard, or none at all, one may be reasonably certain that part of the charge has dissipated through the insulation. Faulty condensers should be replaced with new ones of the same capacity and make. If a condenser is entirely short-circuited it may be impossible to get any output voltages across two or more terminals, as there is then a complete path for the positive current emerging from the choke coils to return to the negative line, which is the path of least resistance. If, however, the condenser has a leaky insulation through which the charge gradually dissipates, it has lost its function to supply energy for an instantaneous overload, and in addition to this considerable fuzziness and distortion will often be evident. A short-circuited buffer condenser may cause a noticeable hum. At other times a crackling noise will be evident in the loudspeaker, resulting from the same cause. A new condenser is the remedy. Details on testing condensers are shown in Fig. 210.

Faulty Resistors.—Faulty resistors, which do not stand up under a heavy current and voltage, may cause crackling noises resembling static, or complete inoperation of the receiver. This is the case when they do not permit enough variation of the voltage. When the detector and radio-frequency resistors, and sometimes even the amplifier resistors, become hot, examine them to see whether part of their winding touches the metal case of the eliminator, or whether the resistor is of the correct capacity, a check which is necessary in case of homemade eliminators. In

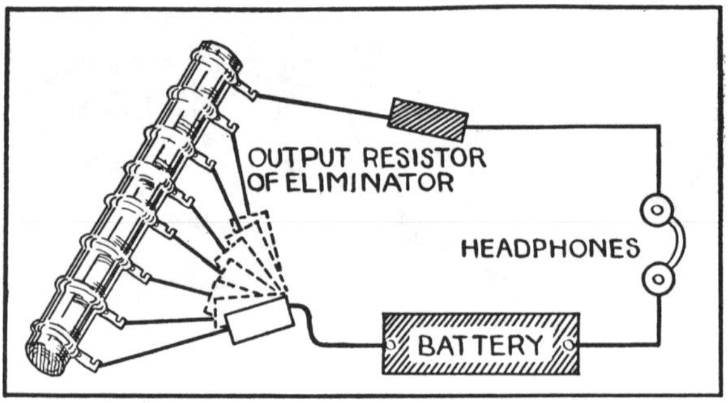

Fig. 125. Output Resistors Sometimes Develop Open Circuits, Which Can Be Detected by Means of the Headphone Tester

many eliminators employing a number of fixed resistors, in addition to the variable ones, it is possible that a variable resistor becomes short-circuited for some reason or other, and then a fixed resistor, in series with it, may be subjected to a greater pressure, with the result that it will become excessively hot. Burned-out or open-circuited elements of resistors will of course cause total inoperation of the receiver. To determine whether or not a resistor is open-circuited, make a headphone test. Touch one testing tip to the B-negative line and the other to the side of the resistor not connected to the B-positive line. If no click is heard, one can be certain that the winding is open-circuited although, perhaps, if another resistor, either fixed or variable, is connected in series to the one being tested, the trouble may be found in this one, which should be tested in the same way. There is one objection to the use of variable resistors in eliminators, namely, that it is difficult for the average fan to adjust them correctly in order to get the proper plate voltages for the tubes. For this reason some manufacturers are making use of fixed resistors having a number of taps, these being designated as unit strips. A definite voltage can be obtained from each one of the taps. The method of testing a unit-strip resistor is shown in Fig. 125.

Obtaining C-Supply from Eliminator

If desired, C-voltage can readily be obtained from a B-eliminator, provided the eliminator is of good design, and has a filter that effectively smooths out the pulsations of the current. However, it must be remembered that there are very few eliminators that are really suitable for providing C-voltage. The prime

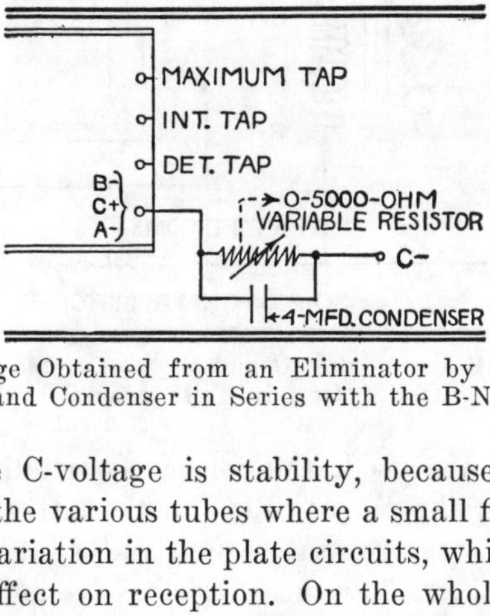

Fig. 126. C-Voltage Obtained from an Eliminator by Connecting a Variable Resistor and Condenser in Series with the B-Negative Output

requisite of the C-voltage is stability, because it controls the grid circuits of the various tubes where a small fluctuation means a considerable variation in the plate circuits, which of course, has a pronounced effect on reception. On the whole, it is better to use C-batteries; they are dependable, as they produce a steady direct-current, their cost is very little, and they last from six months to a year in the average receiver.

But if a B-eliminator has extremely good filtration, changes to provide C-voltage can be made as indicated in Fig. 126. The B-negative output of the eliminator is connected to one side of a 0 to 5,000-ohm variable resistor. The other terminal of this resistor is connected to a C-negative terminal, and a 4-mfd. condenser is shunted around the resistor. If one does not wish to make these internal changes, the method shown in Fig. 127 can be followed, the C-resistor being installed in the receiver. It is connected between the B-negative terminal of the receiver and the line connected to the F-negative terminals of the audio transformers, so that a negative grid bias will be provided for the last two tubes. A 4-mfd. fixed condenser should be shunted across the resistor, as it eliminates coupling which may otherwise occur. If a power tube, such as a UX 171 or CX 371, is used in the last stage, a different arrangement from the one explained above is necessary to provide a grid bias for the last two tubes, for the UX 171 or CX 371 tube takes 180 volts of plate potential and requires a grid-biasing potential of 40.5 volts, while a UX 201A or CX 301A tube in the first audio stage may take a plate voltage of 135 volts, and a grid bias of 9 volts. Thus two separate

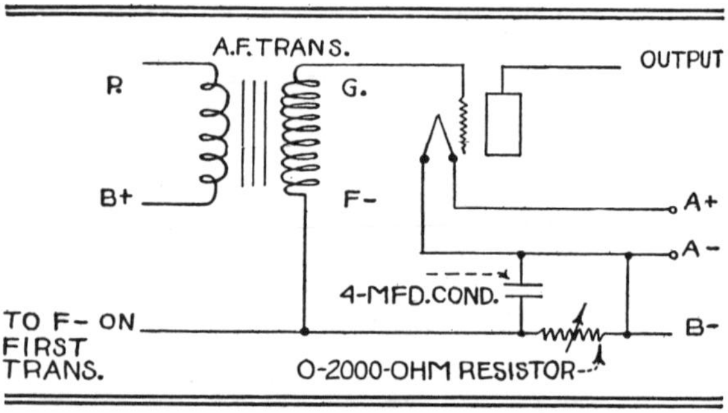

Fig. 127. Variable Resistor and Condenser Connected in the Receiver Circuit to Obtain Drop in Voltage for Grid Bias

variable resistors are used, the former controlling the amount of grid voltage for the power tube and the latter controlling the amount of grid voltage for the second audio tube. The diagram shown in Fig. 128 shows how these resistors are hooked up. Again, it is essential that 4-mfd. by-pass condensers be used at the points indicated to prevent possible coupling.

When C-bias is obtained from an eliminator according to the methods described above, it is necessary to connect the A-negative and B-negative wiring of the receiver together if this is not already the case. On many receivers the A-positive is connected to the B-negative, and considerable damage may be done if this is not changed when adapting the eliminator for C-bias. It must be remembered when taking C-supply from an eliminator, that a condition of instability exists, which is difficult to control. When an especially loud note is reproduced in the receiver, much more plate current is passed than is customary. This momentary surge through the B-system will naturally affect the C-voltage, lowering it to some extent and affecting the operation of the tubes.

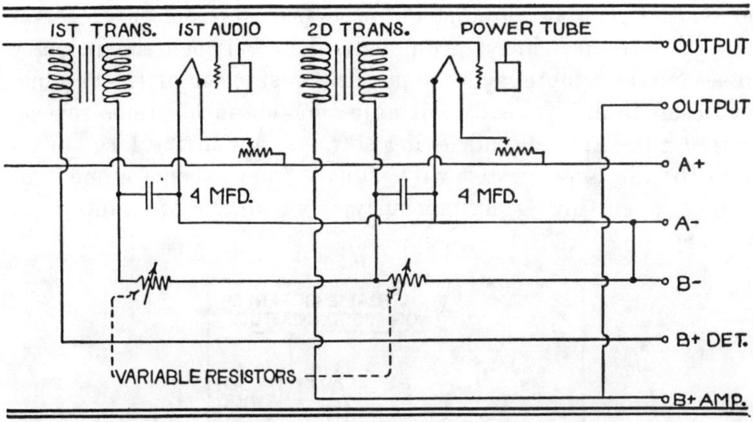

Fig. 128. Using Variable Resistors in Series with the Secondary Windings of Audio-Frequency Transformers in Order to Adjust the Grid Voltages on the Last Two Audio Tubes

Antenna Effect of Battery Eliminators

Broad Tuning a Symptom of Eliminator Antenna Effect.—A radio owner may have noticed, after he has installed a battery eliminator, that the receiver has lost its original sharpness of tuning and has become unaccountably broad—much more so than it was when operated on batteries. A dealer perhaps tells him that this trouble is due to a low A-battery, or is the fault of a nearby broadcasting station, etc. But the owner finds that his A-battery is in good condition, and none of his neighbors complain about the broadcasting station being off its waveband. The real cause of the trouble may be the antenna effect of the eliminator. The house-lighting circuit picks up the signals just as well as any outside antenna, and the signals are passed on to the primary winding and are conducted over both B-negative and B-positive lines to the receiver. The B-negative line is always connected to either the A-negative or the A-positive line, and the A-negative is in turn usually connected to the tuning inductance or ground. Hence all the signals picked up by the house-lighting circuit are transferred to the tuning inductance, and the receiver then has a double aerial system, the excessive inductance of which would naturally make it tune broad.

The remedies are to insert a small fixed condenser in the tuning circuit to overcome the effect of the long antenna system, to shorten the regular antenna or disconnect it from the receiver entirely, or to provide a grounded shield between the primary and secondary windings of the power transformer of the eliminator. Broad tuning is really not objectionable in locations several hundred miles from broadcasting stations, but in the large cities where the air is congested with signals from a large number of stations operating simultaneously, sharpness of tuning is necessary.

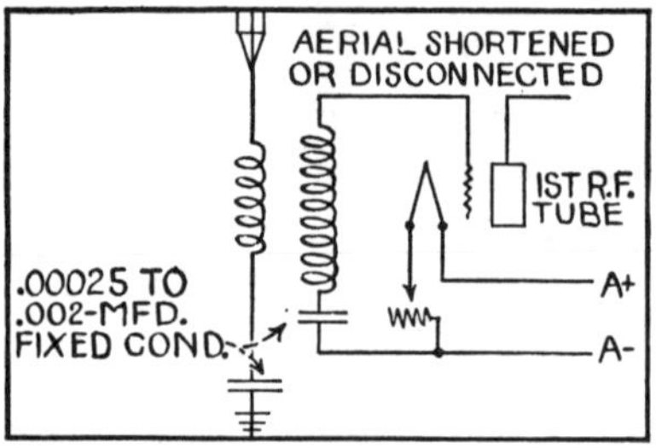

Fig. 129. Small Fixed Condensers in a Two-Circuit Tuning Unit of Receiver to Prevent Antenna Effect of the Eliminator

Methods of Reducing Antenna Effect of Eliminator.—In case one is annoyed by this trouble due to the eliminator, carefully examine the wiring of the receiver and find out just where the A-negative line connects to the aerial circuit. In case of a two-circuit tuner, it may be a direct connection to the secondary winding of the tuning inductance, as shown in Fig. 129. When the connection has been found, cut the wire at a point between the connection and the coil. The two ends formed by cutting the wire are then connected to the terminals of a small fixed condenser ranging between .00025 to .002 mfd. capacity, the most suitable condenser for this purpose being found by experiment. This will have practically the same effect as sharpening the aerial by inserting a small fixed condenser in series with the aerial and the aerial binding post on the receiver. Fig. 130 shows how the condenser is hooked up in case of a single-circuit tuner. Another fixed condenser in the ground line, as indicated in both diagrams, will be still more effective. In some cases it may even be necessary to reduce the length of the regular aerial or disconnect it entirely. At other times it may be advisable to provide a counterpoise. After disconnecting the regular aerial and ground, the counterpoise is connected to the ground binding post of the receiver. This gives a slight condenser effect between the counterpoise and the house-lighting circuit and tends to increase the selectivity of the receiver. Sometimes it will be found quite effective to remove the eliminator from the receiver, as the signal can also be transferred to the receiver by induction.

If it is possible to separate the primary and secondary windings of the power transformer of the eliminator without injuring them it is a good plan to provide a tinfoil shield between the wind-

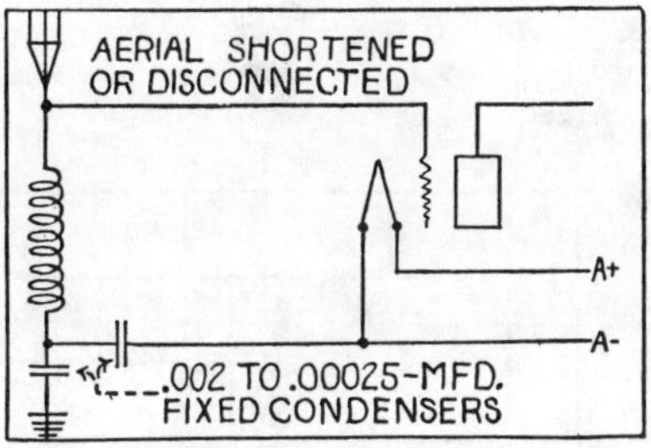

Fig. 130. Condensers in Tuning Circuit of a Single-Circuit Tuning Unit to Prevent Antenna Effect of Eliminator

ings and connect it to the ground circuit. This method of eliminating the trouble is more or less difficult, as most transformers are of the enclosed type and the space between the shell and the windings is filled with an insulating compound. It will often be found helpful, in addition to the methods just explained, to provide a radio-frequency choke coil in series with each output line of the eliminator, with suitable by-pass condensers shunted across the leads, as shown in detail 2 of Fig. 121. This will tend to prevent the radio-frequency currents from getting into the plate circuits of the tubes.

Eliminator Troubles Due to Unstable A. C. Supply

Effects of Unstable A. C. Supply in Receiver.—The voltage of all house-lighting circuits fluctuates; it rises and falls in accordance with the load on the whole distribution system. This gives rise to trouble when using battery eliminators, for the output voltage of the eliminators will fluctuate in synchronism with that of the lighting circuit "Motorboating" is a common trouble which may be set up by such voltage fluctuations. Uncontrollable oscillation of regenerative receivers is another direct result, for receivers of this type operate best just below the oscillation point, and a voltage change may quickly throw them into oscillation. Instability of tuned radio-frequency receivers can also be traced to this cause.

Installing "Glow" Tube and "Ballast" Tube.—The remedy for the trouble resulting from unstable a.c. supply consists in providing some method of stabilizing the output voltage of the elimina-

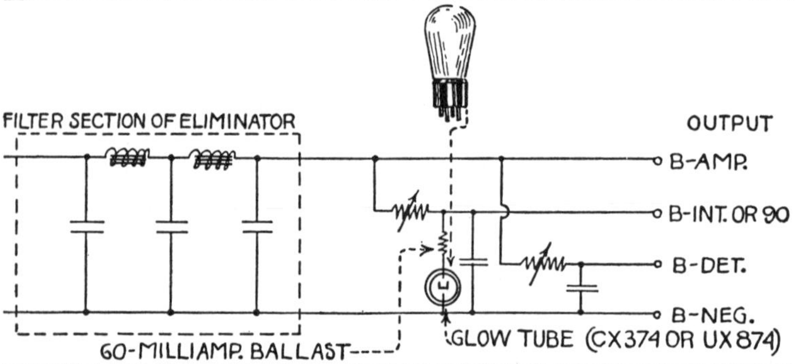

Fig. 131. Method of Connecting a Glow Tube and a Small Ballast Between the B-Negative and B-Intermediate Output of Eliminator to Control the Fluctuations of Voltage Applied to the R. F. Tubes

tor and, if possible, employing some means of controlling the input voltage. A glow tube, such as a UX 874 or CX 374, is especially designed for holding constant the output voltage of an eliminator. Such a tube is connected across the B-negative and B-intermediate or radio-frequency terminals. When installing a glow tube it is necessary to connect it in series with a ballast to limit its current flow to 60 milliamperes. If a glow tube is used in an eliminator, the customary procedure of turning the receiver on first and then the eliminator is reversed because it is necessary to apply 125 volts to the tube elements momentarily to start the glow discharge, but this is done automatically by switching the eliminator on before the receiver. As soon as the receiver is turned on, the voltage applied to the glow tube drops to 90 volts, which is the correct potential. Be sure to connect the glow tube correctly, for it will not function otherwise. If properly connected the tube will show a pink or blue color around the cathode or large circular plate. If the connections to the tube are reversed, a glow will be evident at the small anode terminal. No serious damage to the tube will result in case its connections are temporarily reversed. By using such a glow tube the plate potential on the radio-frequency tubes will not vary more than 2 to 4 volts, even though the line fluctuations are much greater than this. The correct method of installing a glow tube in an eliminator circuit is shown in Fig. 131.

More stable operation of the eliminator can also be obtained by the provision of a "ballast" tube, such as a UX 876 or CX 376. It is connected in series with the primary of the transformer, as

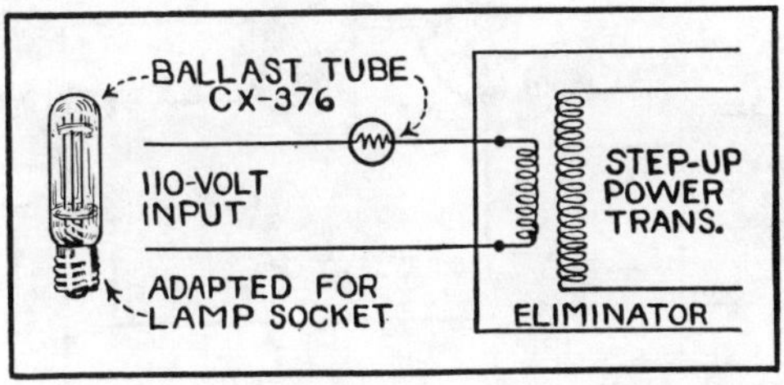

Fig. 132. Ballast Tube Connected in Series with the Power Transformer of an Eliminator to Stabilize Input Voltage

shown in Fig. 132. This tube passes a constant current under fluctuations of voltage and passes the same amount of current through the transformer at all times, which tends to stabilize it.

Methods of Installing Eliminators

Trouble Resulting from Inconvenient Installation.—With the substitution of a B-eliminator for B-batteries, one of the troubles encountered by the average fan is that after the evening's entertainment is over, and the receiver has been turned off, the 110-volt current to the eliminator has been forgotten, with the result that it is left on overnight or perhaps for several days. This of course greatly decreases the life of the rectifying device, making an early replacement necessary. The best method of preventing the trouble is to provide a switch, which will cut off the 110-volt supply at the same time that the filament current to the tubes is turned off, so that one may be certain that the eliminator is not working after the receiver has been shut off.

Hookup for B-Eliminator and 2 to 5-Ampere Charger.—If the fan has a regular charger of any type, of from 1 to 5 amperes capacity, which he uses only at intervals to charge the storage battery, the hookup shown in Fig. 133 will be found most suitable for his needs. One double-pole, double-throw (d.p.d.t.) knife switch and one double-pole, single-throw (d.p.s.t.) knife switch are needed, and these can be purchased at almost any hardware store or electric shop. The former is installed between the charger and the battery according to the diagram. This method of connecting the battery, charger, and receiver makes it impossible for the charger to damage the tubes of the receiver, as the receiver

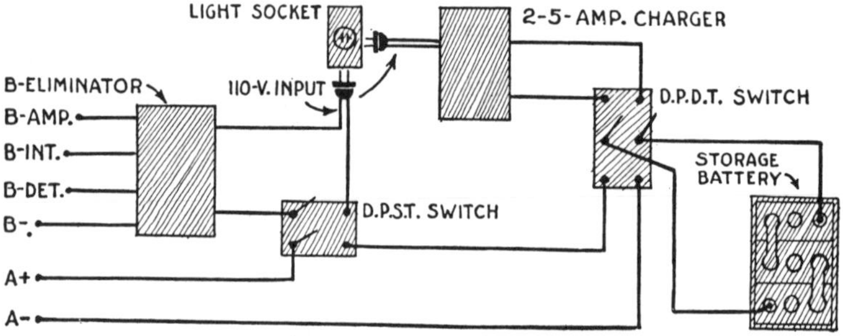

Fig. 133. Hookup for Storage Battery, Charger, and B-Eliminator Using Two Double-Pole, Double-Throw Knife Switches, One of Which Controls the Input of the Eliminator and the Filament Current of the Receiver

is disconnected from the battery the moment the charger is connected, and vice versa. The double-pole, single-throw switch must be of the heavy-duty type suitable for 110-volt current. One of the supply lines from the eliminator to the plug that screws into the lamp socket is cut and the ends connected to the switch at points A and B. Similarly, one of the filament leads to the receiver is cut, and the ends are connected to the remaining two terminals of the switch. Throwing the switch in sets the B-eliminator in operation and at the same time lights up the tubes, the regular filament switch on the receiver being kept on at all times. This hookup has the additional advantage of making the receiver absolutely "dead" when the switch is thrown out, so that if any repairs or adjustments have to be made in the receiver, there is no danger whatever of blowing any tubes or damaging the eliminator.

Hookup for B-Eliminator and Trickle Charger.—Fig. 134 shows a hookup which is convenient in cases where a B-eliminator, trickle charger, and storage battery are used. The usual trouble experienced with an installation comprising a B-eliminator and trickle charger is that, besides letting the eliminator run while the receiver is not in use, as mentioned above, the trickle charger is disconnected from the battery during reception, and one is apt to forget to connect it again afterwards. This may result in a run-down battery, necessitating a trip to a battery-charging station, for it would take three or four weeks to recharge the battery by means of a slow trickle charger. The switch used in this wiring hookup is a triple-pole, double-throw (t.p.d.t.) knife switch of the large type. Follow the diagram closely in making the connections.

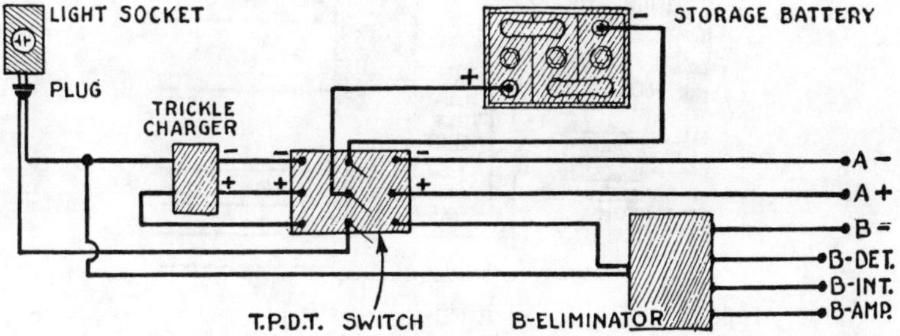

Fig. 134. Triple-Pole, Double-Throw Knife Switch Turns off Trickle Charger, Turns on Eliminator, and Connects A-Battery to Receiver in One Operation, and Vice Versa

By throwing the blade of the switch over to one side, the battery is connected to the receiver, the B-eliminator is switched on, and the trickle charger is turned off and disconnected from the battery. By throwing the switch over to the other side, the receiver is disconnected from the battery, the B-eliminator is switched off, the trickle charger is turned on and is connected to the battery. Also in this case the filament switch of the receiver should be left on all the time.

Hookup for Separate A and B-Eliminators.—Fig. 135 shows a diagram of a wiring hookup used for separate A and B-eliminators. All that is necessary in this case is a 110-volt switch of any kind, which controls the 110-volt input current to both devices, the plug being left in the light socket at all times. In case the light socket is of the key or pull-chain type, or is controlled by a wall switch, a double socket may be used, and the plugs of both A and B-eliminators are connected to it. The switch on the socket is then used to turn the eliminators on and off. With this arrangement also be sure to keep the receiver switch in the "on" position.

Installing Eliminator in Basement.—If for any reason it is desired to install the eliminator in the basement, it is advisable to extend the wires so that the switch controlling its input current can be operated from a position near the receiver. The leads from the eliminator and battery to the receiver must also be extended. They can be passed through the floor or run up through the wall to the receiver, where they can be attached to binding posts mounted on a bakelite panel, which in turn is attached to the wall or to the back of the receiver cabinet, a switch being incor-

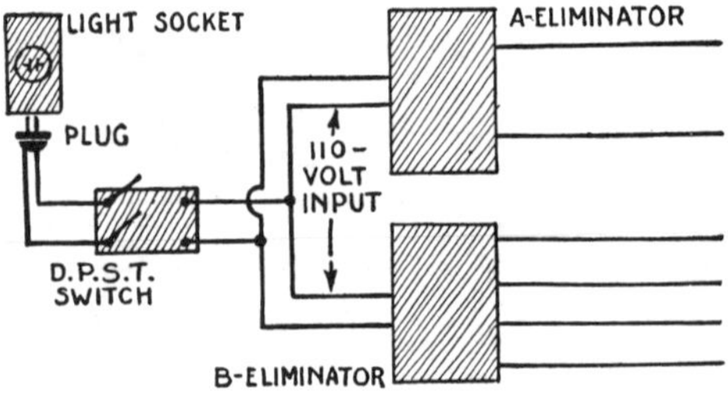

Fig. 135. Input of Both A and B-Eliminators Controlled by Means of Any Ordinary Switch

porated with it or placed near it. A still better method of making such an installation is shown in Fig. 80.

Automatic Switch for B-Eliminators and Charger.—There are a number of devices on the market that shut off the B-eliminator automatically when the filament current to the receiver is shut off. These devices work on the relay principle, the filament-current flowing through a magnet. The magnet, when energized by the filament current, pulls over an armature. This action closes the 110-volt circuit to the B-eliminator and permits it to function. As soon, however, as the filament current is shut off by the switch on the receiver, the magnet is de-energized and the armature is pulled back by means of a small spring. When this happens, the 110-volt circuit is broken automatically and the B-eliminator ceases to operate. A device of this kind is shown in Fig. 136.

These automatic switches are very convenient and commendable, but if the fan does not feel justified in purchasing one, he can arrange a switch to serve the same purpose, but it will not, of course, be automatic in its operation. Methods of doing this are fully explained and illustrated in previous paragraphs.

Fig. 136. Automatic Switch Turns off the Eliminator and Turns on the Trickle Charger When the Receiver Switch Is Opened

Bringing an Old Receiver Up-to-Date with Eliminator Equipment

Many radio fans who have receivers which have become more or less obsolete due to the rapid development of eliminator equipment can nevertheless do much to bring them up-to-date, by substituting a B-eliminator or a power pack, the latter being used in case it is desired to improve the tone quality of the receiver, and to increase its volume and distance-getting qualities. A power pack consists of a B-eliminator and a stage of audio-frequency amplification, making use of a power tube, usually combined in one unit. Fig. 137 shows a typical power pack which has a UX 216B tube for a rectifier and a UX 210 tube as a power amplifier. Obviously, it is not advisable to install a power tube in a receiver without also providing a B-eliminator instead of batteries, for some power tubes pass as much as 100 milliamperes at a zero grid potential, and it is evident that ordinary B-batteries would soon be exhausted when used for furnishing plate current for such a power tube, besides the other regular tubes of the receiver. Power tubes have the ability to amplify signals over a much greater range of frequencies than ordinary tubes, which enables them to reproduce both the high and the low notes of the scale with less distortion than the others.

The power-pack arrangement, therefore, is better than just a B-eliminator only, when it is desired to make a real improvement in an old receiver. However, in case a power pack is used, do not attempt to connect it to the output of a second audio tube, for power packs are designed to operate either directly from the output of the detector or the first audio tube.

The second step toward modernizing the receiver is to provide an A-eliminator. This sometimes necessitates changing the wiring of the receiver to some extent, so that all the tubes will be in series. There are other devices, however, which are essentially filters, and can be hooked up to any ordinary battery

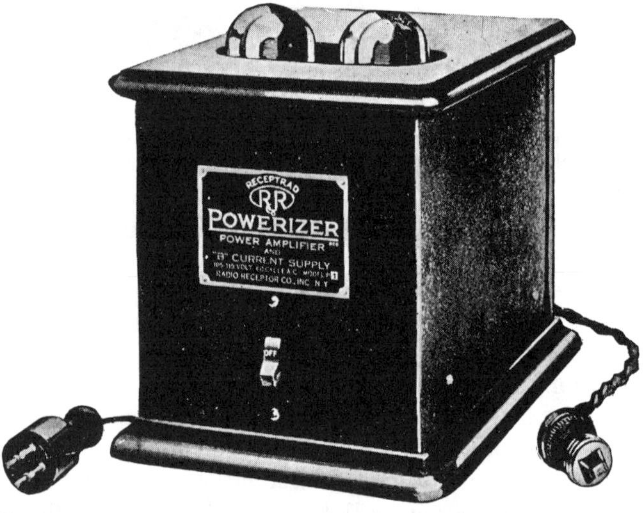

Fig. 137. Power Unit Which Furnishes a Stage of Power Amplification and the B-Supply for the Receiver

charger to supply filament current for the receiver, and it is not necessary, when using these, to change the wiring of the receiver. Either electrolytic chargers or those using a rectifying tube can be employed for this purpose, but not chargers of the mechanical vibrating type. Full directions for installing such a device are provided by the manufacturer and should be carefully followed. A device of this type, containing an electrolytic condenser of large capacity, which makes possible the filtration of so large amount of current as needed for the filaments of the large 5-volt tubes, is shown in Fig. 138.

Eliminators Designed Only for Definite Line Conditions

Certain pieces of electrical apparatus, such as lamps, will work equally as well on a 25-cycle a.c. supply as they will on a 60-cycle a.c. supply. But other appliances, such as motors and transformers, are designed especially to be operated on either a 25-cycle or a 60-cycle a.c. current, but not on both. Therefore such appliances work only when connected to the current for which they are designed. This is also the case with battery eliminators. There are just three points to remember, namely, first, the device should be designed for the voltage of the lighting circuit, which is usually 110 volts; in the second place, the eliminator must be designed for the nature of the lighting current—either alternating or direct, and in the third place, it must be designed for the same number of cycles as the lighting current.

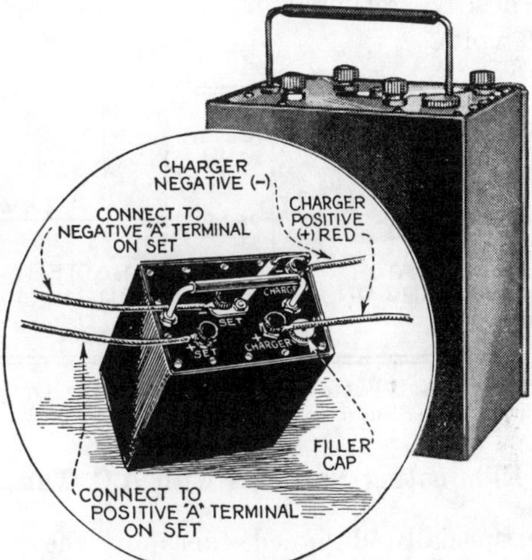

Fig. 138. Large-Size Filter Unit Which Is Connected Directly to the Charger to Provide Filament Current for the Receiver

(Courtesy of "Popular Mechanics Magazine")

Eliminators Operating on Direct Current

Although the great majority of towns are supplied with a.c., there are still many towns and parts of large cities that are provided with d.c. The regular type of battery eliminators designed to operate on a.c. cannot be used on d.c., and specially designed d.c. battery eliminators must be used. With this type of eliminator no step-up transformer is used, and as a rule no voltage higher than 110 volts can be obtained. If higher voltages are necessary for the operation of the receiver, ordinary B-batteries can be connected in series with the eliminator, as shown in Fig. 139. Thus if a potential of approximately 135 volts is desired, a 22½-volt B-battery is obtained, and the negative terminal of the battery is connected to the B-positive amplifier terminal of the eliminator. This arrangement will provide 132½ volts at the positive terminal of the battery, if the eliminator produces 110 volts.

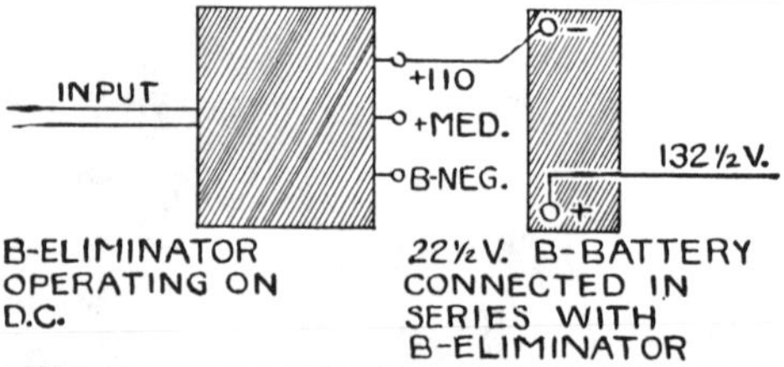

Fig. 139. Connecting a B-Battery in Series with the Output of a Direct-Current B-Eliminator to Obtain Higher Voltage

Eliminating Batteries with A.C. Tubes

There are specially designed tubes on the market, which operate on alternating current supplied by a small step-down transformer connected to the 110-volt house-lighting circuit. These tubes obviate the use of an A-battery, but require plate potentials provided by B-eliminators or B-batteries, and the grid of these tubes must be biased by voltage obtained from a proper C-supply. The exact details of such a.c. tubes and their operation is fully covered in the section on a.c. tubes.

Motor-Generators Eliminate Batteries

There are on the market motor-generator outfits which are designed to supply both A and B current for receivers and thus eliminate all batteries. These devices are made to operate on a.c. or d.c., and can be conveniently installed in the lower compartment of a console, as shown in Fig. 140, or in the basement, if this is desired. A slightly different type is shown in Fig. 141. This device is also a motor-generator, but it furnishes B-supply only, and is run by means of a 6-volt storage battery. It delivers 25 milliamperes of B-current at 120 volts. The drain on the storage battery is only 1.3 ampere, which is slightly more than the amount required to operate four standard UX 201A tubes. A metal case shields the device from causing interference,

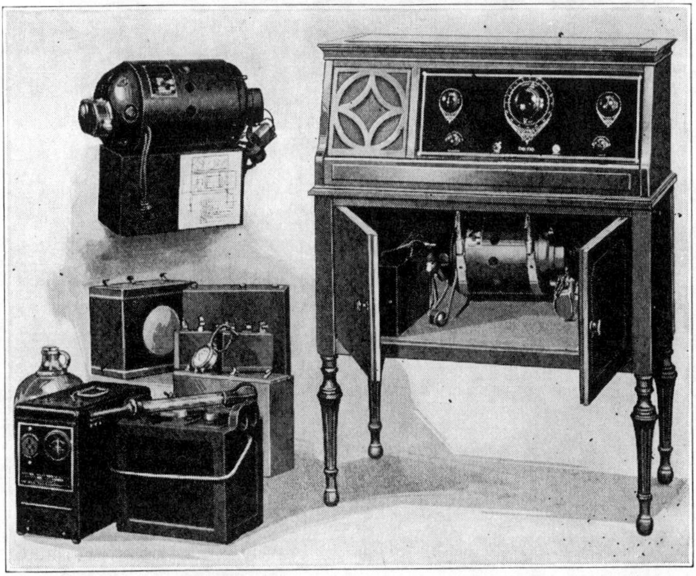

Fig. 140. Motor-Generator Which Furnishes A, B, and C-Supply for a Receiver. It Fits Compactly in the Lower Compartment of a Console and Is Operated from the House-Lighting Current. Left: Equipment Which Is Eliminated by Its Use

(Courtesy of "Popular Mechanics Magazine")

and rubber feet are provided to absorb vibration. Two larger types are also available, delivering 300 and 500 volts of B-potential respectively.

Pseudo A, B and C-Eliminators

There are a number of so-called "eliminators" on the market, which consist of a B-eliminator, a trickle charger, and a small-size storage battery, all in one container. Although they are true eliminators in respect to furnishing B-supply, they are really no eliminators with respect to furnishing A-supply. These power devices are indeed commendable for their compactness,

Fig. 141. Motor-Generator Outfit Driven by a Storage Battery and Used Where House-Lighting Current Is not Available

(Courtesy of "Popular Mechanics Magazine")

simplicity and durability, but they nevertheless share most of the common troubles of storage batteries and chargers, which have been explained in a previous chapter.

CHAPTER VI

TUBE TROUBLES AND THEIR REMEDIES

Efficient operation of radio receivers depends to a great extent on the kind of tubes used and their condition. Many owners are of the opinion that a radio vacuum tube is like an incandescent lamp, and remains in good condition as long as its filament lights. Vacuum tubes are much more complicated than incandescent lamps, and even if they light up perfectly, they may still have defects that will prevent them from operating properly, or even from operating at all. As tubes may cause considerable trouble, the radio owner should provide himself with the best tubes he can obtain in order to avoid this. Using cheap tubes is inviting trouble, and the loud advertisements of many "bootleg" tubes, promising superior tone quality and increased volume and range, should be discredited.

How Vacuum Tubes Operate

A radio vacuum tube consists of a gas-filled glass bulb containing three elements: a filament, a grid, and a plate. Each of

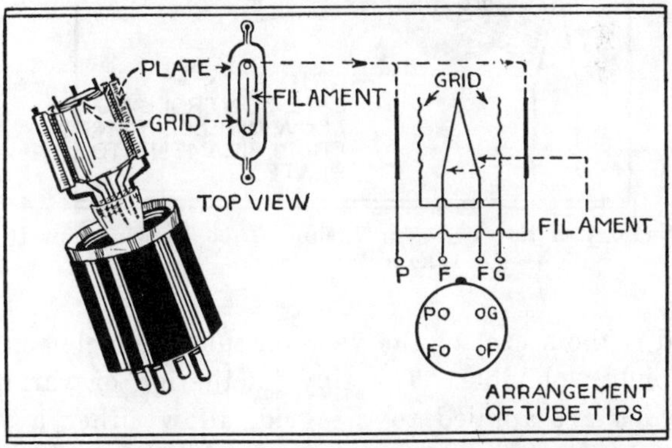

Fig. 142. The Three Elements of a Radio Vacuum Tube Showing how They Are Arranged and Connected Internally

these elements is connected to a tip extending from the base, the filament having two tips, as shown in Fig. 142. In operation, the plate element of a tube is charged with a high potential from a source of direct current, such as a battery, or a socket-power device, this supply being called the B-supply. The filament circuit of a receiver is connected to the negative side of the B-supply, so that a minute current, called the plate current, will flow in the vacuum between the filament and the plate when the filament is heated, which is done by means of an A-supply. The exact rate of the plate current flow depends on several factors: the voltage of the B-supply, the amount of heat generated by the filament, the electron-emitting capacity of the filament, and the amount of voltage impressed on the third element of the tube—the grid. This is a small screen which is interposed between the filament and the plate elements.

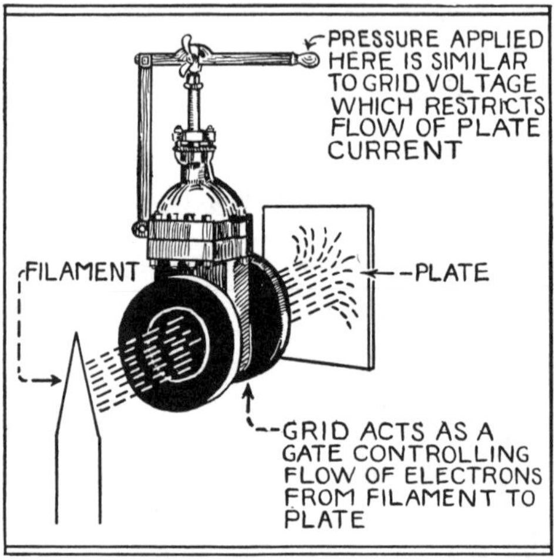

Fig. 143. Analogy of the Grid of a Vacuum Tube Showing how It Operates Like a Gate Valve

The grid is the secret of the vacuum tube; this element makes radio telephony possible. The tiny fluctuation or variations of voltage which are applied to the grid, allow either a small or a large amount of plate current to flow between the filament and the plate, and the grid thus acts as a valve or gate as shown in Fig. 143. A negative voltage on the grid restricts plate current, and a positive voltage allows a considerable amount to pass. Thus a small grid voltage, such as is picked up by the aerial of a receiving set, controls a comparatively large current in the tube, and in this way the minute fluctuating radio-frequency waves can be amplified many times. After the first tube in a receiver has amplified the original radio-frequency signal, the output of the tube is applied to the grid of the next tube, which amplifies the signal still more. This is radio-frequency amplification. A detector then transforms the inaudible trains of radio-frequency vibrations to lower frequencies, which are passed through one or more vacuum tubes and a reproducer, so that they become audible to the human ear.

Definite Operating Conditions of Tubes

Tubes are specially designed for definite operating conditions. For instance, the filament may be of a 1½, a 3, or a 5-volt type, and the tube is constructed to permit certain plate and grid voltages, which can be varied within limits. These voltages must correspond to each other in certain proportions and according

Fig. 144. Rheostats on Receivers Should not Be Turned up too High, or the Tubes Will Become Paralyzed

(Courtesy of "Popular Mechanics Magazine")

to the design of the tube. Usually the plate and grid voltages of a tube or set of tubes in a receiver are fixed, especially in case B-batteries are used, but the filament voltage is variable, being controlled by means of a rheostat.

Use of Rheostats.—A rheostat consists of a resistance-wire winding and a movable slider which can be rotated by a knob in order to get any variation of resistance desired. Rheostats are inserted in series with the tubes, so that all the filament current must pass through them. The average radio fan assumes that the rheostat must be operated for volume and therefore tends to advance it too far, permitting too much current to pass through the tubes. This practice, however, shortens the life of the tubes considerably, for they will soon lose their electron-emitting capacity when their filament is burned too high. This can be avoided by not turning the rheostats so high, as shown in Fig. 144. If a tube has been overloaded by burning it too high, it can sometimes be restored to its normal operation by reactivation, which will be explained later. However, reactivation will not always remedy the condition, and if this is the case, the tube must be replaced with a new one. Too low a filament temperature, if continued for any length of time, is also injurious to the filament, as it then becomes brittle and may break under vibration.

Automatic Filament Control.—To obviate these troubles the filament should be operated at the correct temperature at all times. The substitution of Amperites, which, it is claimed, control the tube filaments automatically, for the rheostats, will be found advantageous. These resistance devices consist of a special alloy, which is chemically coated and works on the thermoelectric principle, its resistance being adjusted automatically according to the requirements of the tube. The use and installation of Amperites is fully covered under the section dealing with fixed resistors.

Necessity of Proper Grid Biasing.—By grid biasing is understood the voltage applied to the grid of a tube to hold it either negative or positive, which makes the tube stable in its operation and helps to prevent it from breaking into oscillation. If no separate battery is used for this purpose, the biasing voltage for the grid is obtained from the filament circuit.

The exact amount of grid voltage necessary for certain tubes depends to some extent on the amount of voltage applied to the plate. The higher the plate voltage applied to a tube, the higher must be the grid voltage to get a balanced bias. Grid voltage required for different tubes, at various plate potentials, is given in Fig. 153.

It should be remembered that the lower the negative grid voltage on the audio tubes, the lower will be the internal impedance of the tube, which is the resistance to the passage of current between the filament and the plate. A comparatively low impedance in a tube means better tone quality in the receiver. On the other hand, a high grid voltage increases the internal impedance of a tube and, if excessively high, harsh-toned and distorted reception may result. A high grid voltage means less consumption of plate current, and this in turn assures a longer life of the B-batteries than is the case with a low grid bias.

Adjusting Grid Bias.—The most satisfactory method of adjusting the grid bias is to insert a 0 to 10-reading milliammeter in the plate line to the tube that is to be adjusted, as shown in Fig. 111. The plate and grid voltages are then adjusted so that when the receiver is set in operation the milliammeter needle will not fluctuate, even when strong signals are reproduced. If the needle of the milliammeter deflects toward zero when a specially strong signal is reproduced, the C-voltage on the grid is too high, and if it deflects in the opposite direction the C-voltage is too low. When the grid and plate voltages are correct, there should be no appreciable fluctuation of the needle, when a station is tuned in or out. Ordinarily, it will be found that the use of correct C-voltages as recommended by manufacturers will give satisfactory results. However, to adjust the C-voltage accurately, connect a variable resistance in series with the C-battery and the grid of the tube.

Life of Tubes

All tubes depreciate with use, and their life is usually estimated in the number of hours that they will give good service under normal operating conditions. Although the average life of tubes may be estimated, manufacturers do not, as a rule, guarantee them for any definite length of time, for the use of incorrect voltage and current will shorten their life considerably. It is claimed that the average life of a UX 201A or CX 301A tube is about 1,500 hours, which represents about two years of service to the average radio owner. However, tubes have been found to last much longer than this, while others give out more quickly.

Paralyzed Tubes Caused by Constant Overloading.—Most tubes have thorium-impregnated filaments, and when tubes are burned too high the surface layer of the thorium is gradually lost and the emission capacity is greatly decreased, which results in weak signals and failure to get distant reception. When this occurs, the rheostats are usually turned up still higher and the tubes, thus overloaded, become paralyzed. In most cases paralyzed tubes can be restored to a great extent by a process called reactivation, but sometimes the paralyzed condition cannot be remedied and the tube must be discarded.

Paralyzed Tubes Due to Accidental High Voltage.—Many radio owners tinker in their receivers with screwdrivers or other metal tools to tighten up loose connections, adjust condensers, etc., without first disconnecting the batteries and removing the tubes. This should never be done, for the tool may accidentally be touched on some metal and a short circuit may then result, which allows the high-voltage B-battery current to get on the filament lines. Tubes are often burned out in this way, but sometimes they can still be lighted afterwards. It will be found, however, that such exposure has paralyzed them considerably, decreasing the volume of reception or causing total inaudibility. If this has happened, take the tubes out and test them. Most likely they will need reactivation, but in many cases this will not improve their condition. Such accidents can be prevented by disconnecting the B-supply while the adjustments are being made inside of the receiver.

Tube Testing

Periodically the tubes of a receiver should be taken to a neighborhood radio dealer, who will usually test them as a matter of courtesy, but, in order to guard against an incorrect report, whether made inadvertently or otherwise, every radio fan should acquaint himself with the meters used for tube testing so that he will know which meter readings indicate good, fair, and poor tube conditions.

Commercial Tube Tester.—Fig. 145 shows a good type of commercial tube tester, which is used extensively by dealers. It has five separate meters which permit one to tell the exact condition of a tube at a glance. This tube tester consists of a case having a bakelite panel on which the various instruments are mounted within plain view of the operator. A 0 to 6-reading voltmeter is set in the upper left-hand corner; it is used to indicate the number of volts applied to the filament of a tube being tested.

The zero-grid voltmeter just below it has a negative and a positive reading from zero to 10 volts, and it is used to determine the polarity and exact voltage of grid bias applied to the tube. The center meter is an 0 to 10-reading milliammeter, which shows the amount of plate current being passed by the tube under test. The lower right-hand voltmeter is graduated from 0 to 120, and indicates the amount of B-battery voltage applied to the plate of the tube. The upper right-hand instrument is a 0 to 1.2-reading ammeter, which shows the amount of current passing through the filament. A tube socket is located between the filament volt-

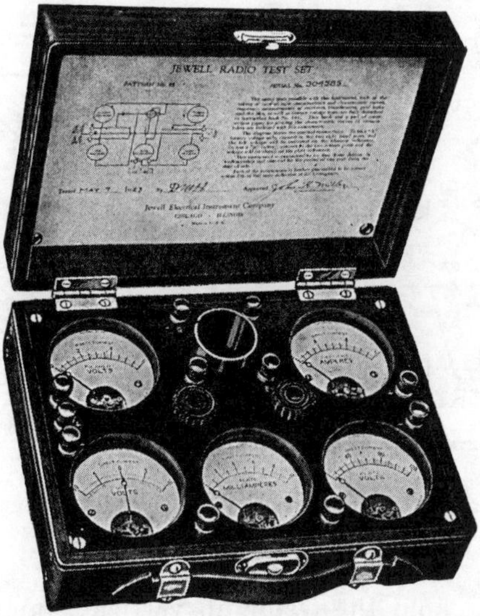

Fig. 145. Commercial Tester with Which the Characteristics of Vacuum Tubes Can Be Determined

meter and the filament ammeter. Immediately below it are two variable controls, the left-hand control being a potentiometer, which is used to vary the grid bias, and the right-hand control a rheostat, used for varying the current and voltage applied to the filament of the tube. Binding posts are provided on three sides so that A, B, and C-batteries can be connected to the tester as indicated in detail 1 of Fig. 146. For testing 5-volt tubes a 6-volt A-battery, a 45-volt B-battery, and two 9-volt C-batteries, connected in series, as shown, are used.

Protecting Milliammeter from Burning Out.—When using a tester of this kind it is a good plan to protect the milliammeter against being burned out by an internal short circuit between the plate and the grid of a tube being tested. Protection against this damage can be had by connecting an ordinary 10-watt tungsten or carbon lamp in one of the B-battery lines, and using a switch so that the lamp can be shunted out if the tube has no short circuit between the grid and plate, which is indicated by the deflection of the needle of the plate voltmeter when a tube is inserted in the tester. The method of hooking up this protective lamp is shown in detail 2 of Fig. 146. After a tube is inserted

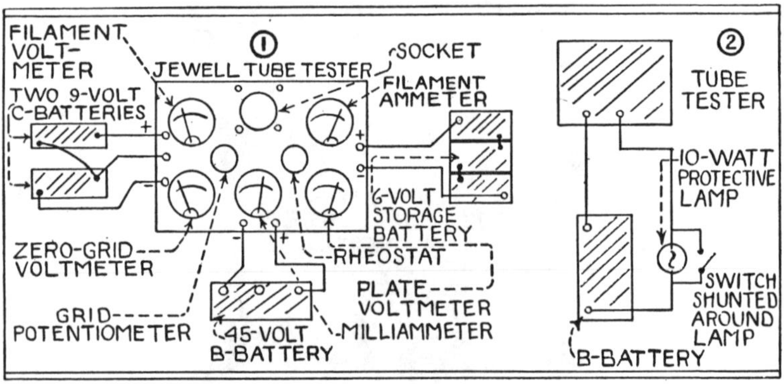

Fig. 146. (1) Diagram Showing Various Meters and Controls of a Jewell Tube Tester and Indicating how the Batteries Are Connected. (2) A Tube Tester Protected by a 10-Watt Lamp in the Plate-Supply Circuit

into the tester, hold the tester up vertically and gently tap the tube, for tubes occasionally have intermittent short circuits, which can be detected only when the tube is held at an angle. There are many other types of tube testers on the market, which are not so elaborate and are therefore cheaper than the one shown in Fig. 145. One of these is shown in Fig. 147, the handy feature of this type being that it has an extension cord provided with a plug, which fits into the sockets of a receiver, from which current is obtained for the tube under test, while the receiver is in operation.

How to Test a Tube.—To test a tube, insert it in the socket of the tester while the protective lamp in the B-circuit is connected, shunting this out if no sudden deflection is noticed in the plate voltmeter. The rheostat is turned up so that a reading is obtained on the filament voltmeter and filament ammeter. The reading to which these meters are adjusted varies with different types of tubes; the cartons in which the tubes come will usually have the filament voltage and current printed on them, and this information is also given in the tube chart, Fig. 153. In case of UX 201A or CX 301A tubes, which are in general use, the filament voltage should be 5 volts, and the filament current .25 ampere. As both the voltage and the amperage are regulated with one control there may be a slight variation from these constants, but this is permissible as tubes are not all exactly alike, which is to be expected in quantity production. If the current

Fig. 147. Convenient Tube Tester Which Is Plugged into the Socket of the Tube Being Tested
(Courtesy of "Popular Mechanics Magazine")

is quite a little below normal, it may be necessary to increase the voltage above normal in order to get proper results when testing.

Read the plate voltmeter to see whether the plate voltage is correct. After the plate voltage has been read, adjust the potentiometer so as to get a zero reading on the milliammeter. Turn the potentiometer still farther until the zero-grid voltmeter also reads zero, after which the potentiometer may be turned up gradually. At every volt graduation of the zero-grid voltmeter, the reading on the milliammeter is noted. For normal UX 201A or CX 301A tubes the readings on the grid voltmeter and on the milliammeter should be somewhat similar to the readings given in the following table. Slight variations are to be expected, as no tubes are exactly alike.

Fil. voltage .... 5 volts
Fil. current .... .25 ampere
Plate voltage .... 45 volts

Grid Voltage Negative	Plate Current in Milliamperes
6	0.0
5	0.1
4	0.2
3	0.4
2	0.7
1	1.1
0	1.6
Positive	
1	2.1
2	2.8
3	3.5
4	4.2
5	5.0
6	5.8
7	6.6

Grid Voltage Positive	Plate Current in Milliamperes
8	7.4
9	8.3
10	9.2

Testing Alkali-Vapor Tubes.—Tubes containing alkali vapor, such as the UX 200A and CX 300A tubes, which are used for detectors, do not always give correct readings as soon as they are inserted in the tester. The formation of the vapor occurs during the first three minutes after the tube is lighted, and during this time the tube characteristics change rapidly, causing variations in the readings of the plate current and performance of the tube while it is in operation in the receiver. As soon as the tube has become warm, in about three minutes, the test readings are uniform and no further changes occur. Readings taken during the period of change are variable and are of no value as a true indication of the tube's characteristics. However, as the change does not usually begin until the tube has been in the tester about one minute, it is possible to make a quick test immediately after it has been placed in the socket; otherwise the readings should be taken at the end of three minutes.

Tests on these tubes should be made on a plate potential of 45 volts and a 4½-volt negative bias on the grid. As they are intended for detector service, a very careful test of these tubes is necessary. Plate-current readings on the milliammeter indicate only the condition of the filament and cannot be depended on as an indication of the tube's sensitivity. Therefore, the best practical test of a tube is to check its performance while it is in actual operation in the receiver on a weak incoming signal, rather than on a strong signal.

A plate-current reading obtained while it is in actual operation in the receiver, with a grid leak and condenser connected, is quite different from that obtained in the tester. In actual service conditions the plate-current reading is approximately 1 milliampere, varying only slightly from this value, whether the grid return is connected to the negative or to the positive filament line. The difference in readings, as compared with those obtained in the tester, is due to the fact that the grid current furnished by the tube is sufficient to establish a voltage drop in the grid leak equivalent to the use of a grid-biasing voltage of 2 or 2½ volts.

When a tube is in actual service a hiss is frequently heard at the time that the changes in plate current, mentioned above, are occurring. If a milliammeter is placed in the detector-plate circuit, it will be noted that the initial reading is perhaps 1 milliampere, remaining at this value during the first minute, while the reading gradually rises to 2 or 3 milliamperes during the second minute. It is during this change of current that the hiss sometimes occurs. The plate current then returns to a value close to that of the initial reading, after which the tube will operate quietly. If the hiss is objectionable, the operator should remove the loudspeaker plug from the horn jack on the receiver for two minutes, so that the hiss will not be heard.

Testing Other Tubes.—When testing other types of tubes, the tester should be adjusted so that the constants will correspond to the figures given on the carton in which the tubes are sold. Also see Fig. 153. Good tubes show a considerable change in plate current on the milliammeter to a small corresponding change in the grid voltage. This general characteristic is equally true of all types of tubes, whether they are used for detection or for amplification. Under test some tubes may appear to have a bluish color internally, which is an indication of excessive plate voltage or of the presence of some air inside the tube. If the color does not disappear after the plate voltage has been reduced, the tube is defective and should be discarded.

Matching Tubes.—It has been found advisable in case of radio-frequency receivers and superheterodynes to provide "matched" tubes for the radio-frequency stages of the former, and for the intermediate stages of the latter. Matched tubes are those having corresponding plate-current readings at various grid voltages. In order to determine this, the tubes are carefully tested and the readings of the milliammeter and the grid voltmeter are taken down at every change of grid bias. A considerable improvement in reception will be evidenced in certain types of sets as mentioned, if matched tubes are used instead of tubes picked at random. When tubes are being matched in the tester the same plate voltage should be used as that which is to be applied to them in the receiver.

Homemade Tube Tester.—A good homemade tube tester using the same instruments and hookup as the one shown in Fig. 145 can be made by any radio fan. Get a substantial wooden box, about 10 in. long, and 7 or 8 in. wide, preferably with a hinged cover. Provide cleats on the inside and obtain a piece of ¼-in.

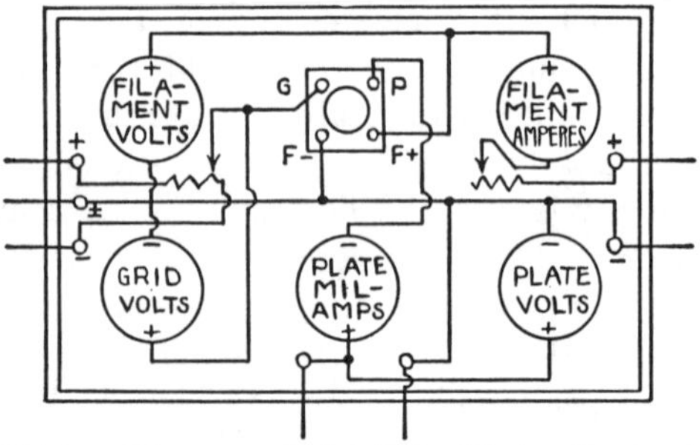

Fig. 148. Wiring Diagram of the Tube Tester Shown in Fig. 145

bakelite panel to fit snugly inside of the box on the cleats, so that the top of the panel will be flush with the edge of the box. Then lay the instruments on the panel symmetrically, and mark off for drilling. After the holes have been drilled for the instruments and for the screws, which are used to attach the panel to the cleats, fasten the instruments securely in their respective positions, and proceed to wire them according to the diagram shown in Fig. 148, using "spaghetti" to insulate the wires, and being sure to solder all connections securely.

Tube Reactivation

When a tube has lost its electron-emitting capacity and is partly or wholly paralyzed so that it passes a very small plate current, it can often be restored to an active condition by a process of reactivation or "boosting." Fig. 149 shows a tube reactivator and how it is used. It should be remembered, however, that only tubes having a thorium-impregnated filament can be reactivated, and fortunately most of the tubes in use are of this type, except the WD 11 or C 11, and WD 12 or CX 12 types, which have oxide-coated filaments. In case of thoriated filaments,

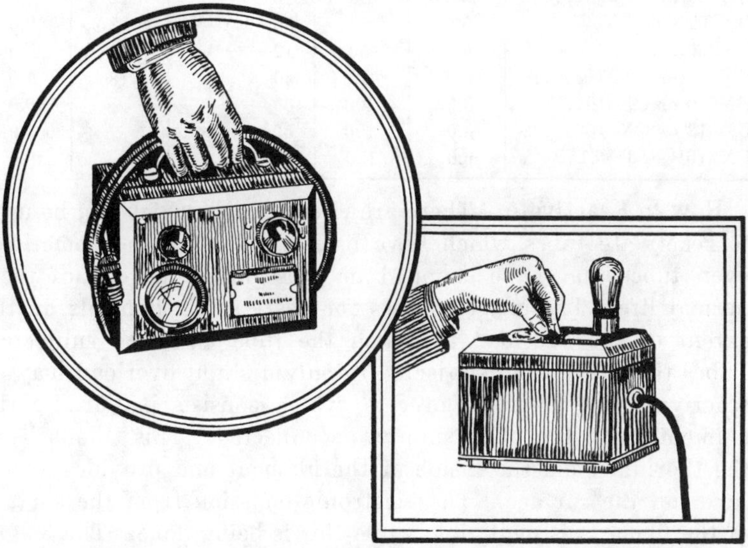

Fig. 149. All Radio Vacuum Tubes Which Have Thorium-Impregnated Filaments Can Be Reactivated by Means of a Tube Reactivator. This One Is Provided with a Milliammeter to Test the Plate Current of the Tube Under Test

the coating of thorium on the surface gradually gives out and causes the tube to lose its sensitiveness. By reactivation the thorium inside of the filament is brought to the surface, thus permitting the tube to function normally again. Since the symptoms exhibited by exhausted tubes and weak B-batteries are practically the same, the owner is apt to ascribe the condition of his set to the latter, and to change the batteries, when it is really the tubes that are at fault.

When to Reactivate Tubes.—If one is not sure that the tubes need reactivation, a simple test of the tube emission or plate-current flow, will show its condition. When the plate current of the UX and CX tubes, given in the following table, drops to the values indicated under "Minimum Emission," reactivation is necessary.

Type of Tube	Fil. Voltage	Grid and Plate Voltage	Minimum Emission in Milliamps.	Maximum Flashing Voltage	Maximum Aging Voltage
UV199 or C299 / UX199 or CX299	3.3	50	6	12	4.0
UX120 or CX220	3.3	50	15	12	4.0
UX201A or CX301A	5.0	50	25	18	7.0
UX200A or CX300A	5.0	50	12	18	7.0
UX171 or CX371	5.0	50	50	18	7.0
UX210 or CX310	6.0	100	100	..	10.0
UX213 or CX313	4.0	100	50 per anode	..	7.0
UX216B or CX316B	6.0	125	100	..	10.0

How to Reactivate.—There are two methods which can be used to reactivate tubes which have been overloaded and sometimes even tubes that have dropped in emission at the end of their normal life. The exact process for best results depends on the extent of the overload to which the tube has been subjected. Tubes that have been subjected to only a slight overload may be reactivated by a simple process, which consists in burning the filament, with the plate supply disconnected. This "boils" out the thorium from the inside of the filament and provides a new layer on the surface. The electronic emission from the surface of the filament is negligible while this is being done. The length of time required to reactivate a tube by this process is usually from ½ to 1½ hours, depending entirely on the condition of the tube. Occasional tests to ascertain the plate emission should be made, and in this way one can readily determine when the tube has been brought back to normal. Any radio owner can reactivate his tubes in this way without the need of special equipment.

Tubes that have been badly overloaded may not improve perceptibly when subjected to this treatment, and it will then be necessary to use another method of reactivation, employing the use of a reactivator. The tube must first be "flashed" from 10 to 20 seconds. This is merely burning the filament at a voltage indicated under the heading of "flashing voltage" in the table given above. Then the filaments must be burned at the voltage given under the head of "aging voltage," for about 30 minutes. Test the tube for emission, and if it is not restored, burn the filament at the aging voltage for about 2 hours. If the emission is not normal after this has been done, the tube is permanently impaired and cannot be reactivated. No plate voltage is ever applied during reactivation. It is advisable to be certain of the exact voltages applied to the filament by means of a suitable voltmeter. A small percentage of tubes that are reactivated will burn out when the flashing voltage is applied.

Reactivation Sometimes Impossible.—Tubes which have internal short circuits between elements cannot be reactivated, and it is advisable to check for such conditions with a C-battery and a pair of phones, as shown in Fig. 157, before attempting to reactivate them. If a click is heard when touching the grid and plate terminals of the tube, or when touching the grid and filament terminals, there is a short circuit, and the tube should be discarded. This test should be made when the tube is not connected in any circuit. Tubes in which the vacuum is impaired cannot be reactivated. This is usually indicated by a filament-current reading above the rated value. If considerable air is present in the tube, the filament will not light up at all unless the filament voltages are raised well above normal, in which case the filament will usually burn out.

"While-You-Wait" Reactivation.—Reactivation of tubes by the "while-you-wait" method is not recommended, for in such cases the work is usually done very quickly, sometimes within 10 minutes, and higher voltages than those given in the table above are used. This procedure materially shortens the life of the tubes, and tubes reactivated in this way soon fall back to their minimum emission value. Besides, the use of higher voltages greatly increases the percentage of tubes burned out.

Reactivation Equipment.—There are several types of tube reactivators on the market. Complete instructions for operating them are provided, so that the work is really simple, enabling anyone to do it. These devices are connected to the house-lighting circuit, which restricts their use to city dwellers. On the farm and in locations where alternating house-lighting current is not available, the owner can use the first method of tube reactivation as explained above. However, it is possible to use the second method provided two 6-volt storage batteries are connected in series, so that a flashing voltage can be obtained. A 6-ohm rheostat is used to control the voltage and a 0 to 50-reading voltmeter is connected, as shown in Fig. 150.

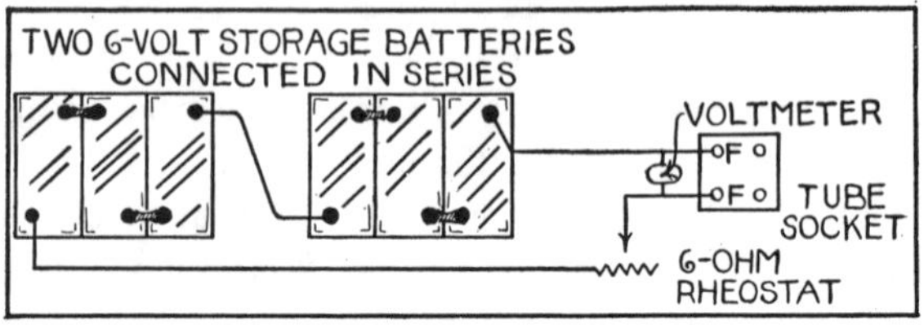

Fib. 150. When no House-Lighting Current Is Available to Operate a Tube Reactivator, Two 6-Volt Storage Batteries, Connected in Series, Can Be Used to Reactivate Tubes

Homemade Tube Reactivator.—A tube reactivator consists of a step-down transformer, a socket, and a voltmeter. The transformer is tapped so that a flashing and an aging voltage can be obtained. However, if the radio owner does not feel justified in purchasing a reactivator, he can make one himself. All the equipment necessary is a toy-train transformer, tapped for various voltages; a 6-ohm radio rheostat, a 0 to 50-reading voltmeter, and an ordinary radio-tube socket. These instruments are neatly

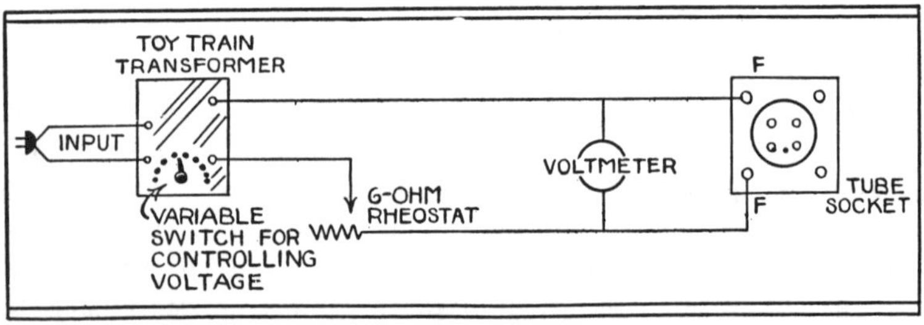

Fig. 151. Homemade Tube Reactivator Consisting of a Toy Train Transformer with a Variable Voltage Range, a 6-Ohm Rheostat, a Voltmeter, and a Vacuum-Tube Socket

fastened to a bakelite panel, attached to a small, wooden box, and are then wired up according to the diagram shown in Fig. 151. Use regular radio bus-bar wire for this purpose, but slip "spaghetti" over it to eliminate danger from short circuits, and be sure to solder all connections so that they will be electrically and mechanically tight.

Frequent Reactivation Proof of Improper Operation.—When operated under normal voltages, the better class of radio tubes do not require reactivation. This is proved by tests made on UX and CX tubes, which were subjected to normal operating voltages for 1,500 hours, which is equal to about two years of service for the average fan. It has been found in such tests that the plate-current value over this period is practically constant, making reactivation unnecessary. Such constant emission is, however, obtained only when the proper voltages are applied to the elements of the tube, and therefore it is advisable to use

Fig. 152. Every Radio Owner Should Have a Voltmeter with Various Ranges for Testing the Voltage of A or B-Batteries or Eliminators
(Courtesy of "Popular Mechanics Magazine")

a voltmeter on the receiver, as shown in Fig. 152, which should be used as a check in order to protect the tubes against overloads. Furthermore, proper adjustment of voltages means a saving in battery current.

If it is necessary to reactivate tubes every month or at frequent intervals, it is proof that the tubes are being overloaded, and in such cases the following precautions should be observed: Do not burn the filament voltages in excess of the rated filament terminal voltage. Keep the filament rheostat set as low as possible. Do not use high plate voltages unless proper grid bias is provided. Without a grid bias, it is best not to exceed a potential of 67½ volts on the plates of the radio-frequency and audio-frequency tubes, while the detector-plate voltage should not exceed 45. Be careful, when changing battery connections, to see that the polarities are not reversed. If the leads connecting the A and C-batteries are reversed, signals may still be heard, but they will be faint or distorted. Such a reversed condition often causes the operator to turn the filament to a high setting, thus injuring the tubes without improving results appreciably. Always check the battery connections after making changes. If an insensitive tube and a sensitive tube are controlled by the same rheostat, there will be a tendency to overload the good tube in order to obtain satisfactory operation from the poor one. The poor tube should be reactivated or replaced.

When using UX 199 or CX 299, UV 199 or C 299, or similar 3-volt tubes, and the owner does not desire to purchase a filament voltmeter, the condition of the tubes can be checked in the following way: Connect two dry cells, which are absolutely fresh, in series and hook them up to the filament lines of the receiver in the proper way. The rheostats are then turned up entirely; if satisfactory reception is obtained, the tubes are in good operating condition.

Various Types of Tubes in General Use

So that he will use the right kind of tube for the right purpose the radio owner should have access to information concerning the most popular tubes which are in general use, and their characteristics. Such information is given in the chart in Fig. 153.

UX and CX Tubes.—Tubes designated by the letters UX before the number are called "Radiotrons" and are manufactured by the Radio Corporation of America, while tubes bearing the letters CX are Cunningham tubes and are sold by the E. T. Cunningham Radio Corp. Each of these types of tubes is very popular, and a number of them are shown in Fig. 154. The UX 201A or CX 301A are ideal tubes for general-purpose work and will serve well in any or all sockets of a receiver designed for these tubes. However, especially in 5 and 6-tube tuned radio-

CHART OF TUBE CHARACTERISTICS (Fig. 153)

Type of Tube	Use	Fila-ment Vol-tage	Filament Current	Detector Plate Voltage	De-tector Plate Current	Amplifier Plate Voltage	Grid Voltage (C-Batt.)	Amplifier Plate Current	Plate Impedance	Amplifi-cation Constant	Mutual Con-ductance
5-Volt Detectors and Amplifiers											
UX-201A or CX-301A..	Det. Amp.	5.	.25	45	1.5	45 67.5 90 135	.5–1.5 1.5–3 4.5 9	.9 1.7 2.0 2.5	18,500 14,000 12,000 11,000	8.	460 600 710 775
UX-200 or CX-300.....	Det.	5.	1.	16–22.5	1.	....	....	....	....	....	....
UX-200A or CX-300A...	Det.	5.	.25	45	1.5	....	....	....	30,000	2.	670
Ceco A...............	Det. Amp.	5.	.25	37–45	.8	90 135	4.5	3.5	6,800	8.5	1,210
Ceco H...............	Det.	5.	.25	67–90	2.6	....	....	....	10,500	14.4	1,250
Ceco K...............	R. F.	5.	.25	....	....	67–140	....	....	....	....	....
Gem 201A.............	Det. Amp.	5.	.25	45	1.0	90 135	4.5 9.	2.8	12,700	8	800
Gem 200..............	Det.	5.	.25	45	2.0	....	....	....	35,000	20	570
Jaeger JX-201A........	Det. Amp.	5.	.25	20–45	1.0	90	4.5	3.1	8,000	7.5	900
Kenrad KR-0201A......	Det. Amp.	5.	.25	20–45	2.2	90	4.5	3.5	11,000	8	730
Magnavox 201A.......	Det. Amp.	5.	.25	22–90	1.0	90 135	4.5 9.	2.5 3.8	8,570	8.	930

CHART OF TUBE CHARACTERISTICS (Fig. 153)—*Continued*

Type of Tube	Use	Filament Voltage	Filament Current	Detector Plate Voltage	Detector Plate Current	Amplifier Plate Voltage	Grid Voltage (C-Batt.)	Amplifier Plate Current	Plate Impedance	Amplification Constant	Mutual Conductance
Magnavox 150D	Det.	5.	.25	22–90	1.2	......	......	......	......	......	......
Perryman RH-201A	Det. Amp.	5.	.25	22–45	1.0	90 135	4.5 9.	3.5 4.5	10,500	8.	760
Perryman PD-200A	Det.	5.	.25	45	1.1	......	......	......	......	......	......
DeForest— DV-5	Amp.	5.	.25	......	......	90 135	3.	2. 3.5	12,000	9.5	800
D-01A	Det. Amp.	5.	.25	22½–45	......	67.5 135	9	......	10,000	......	......
DL-4	R. F.	5.	.25	......	......	67.5 90	......	3.5	7,800	9. 9.	1,150 1,150
DL-2	Det.	5.	.25	16.5–45	......	67.5 90	......	8.5	9,500	7.	800
DL-15	Det.	5.	.25	22.5 45	......	......	pos.	.5	......	......	......
DL-7	Audio	5.	5.	......	......	135	9.	6.	6,500	7.	1,100

Dry-Cell Tubes: Detectors and Amplifiers

Type of Tube	Use	Filament Voltage	Filament Current	Detector Plate Voltage	Detector Plate Current	Amplifier Plate Voltage	Grid Voltage (C-Batt.)	Amplifier Plate Current	Plate Impedance	Amplification Constant	Mutual Conductance
UV-199 or C-299 UX-199 or CX-299	Det. Amp.	3.3	.06	22–45	......	45 67.5 90	.5–1.5 1.5–3 4.5	1. 1.7 2.5	19,500 16,500 15,000	6.25	320 380 415

CHART OF TUBE CHARACTERISTICS (Fig. 153)—*Continued*

WX-12 or CX-12...... WD-11 or C-11........	Det. Amp.	1.1	.25	22.5	......	45 67.5 90	.5 3 4.5	1.1 1.8 2.6	18,000 17,000 16,000	6.2	340 360 380
Ceco B, BX, C........	Det. Amp.	3.	.06	22–45	.4	45 67.5 90	1.5 3 4.5	.7 2.4 2.8	14,800	6.7	450
Duval 199.............	Det. Amp.	3.	.06	22–45	.4	45 67.5 90	1.5 3 4.5	.8 2.3 2.5	15,000	6.	400
Excell 199.............	Det. Amp.	3.	.07	45	......	45 67.5 90	1.5 3 4.5	1. .... 2.6	15,500	5.8	395
Gem 199, X-199.......	Det. Amp.	3.	.06	22–45	.4	45 67.5 90	2.5 3. 4.5	1.8 2.5 3.3	15,500	6.3	400
Jaeger JX-199.........	Det. Amp.	3.	.06	22–45	.4	45 67.5 90	2. 3. 4.5	.7 2.4 2.8	14,000	6.5	410
Perryman 199.........	Det. Amp. Amp.	3.	.06	22–45	.4	45 67.5 90	1.5 2.5 4.5	1.5 2. 2.5	14,000	6.5	400
DL-3 and DL-3A......	Det. Amp.	3.	.07	45 16.5	......	90	4.5	4	1,450	8.	550

CHART OF TUBE CHARACTERISTICS (Fig. 153)—*Continued*

Type of Tube	Use	Filament Voltage	Filament Current	Detector Plate Voltage	Detector Plate Current	Amplifier Plate Voltage	Grid Voltage (C-Batt.)	Amplifier Plate Current	Plate Impedance	Amplification Constant	Mutual Conductance
POWER TUBES											
UX-210 or CX-310	Power Tube and Oscillator	Trans. 7.5	1.25	....	......	250 350 425	18 27 35	12 18 22	5,600 5,100 5,000	7.5 7.6 7.7	1,330 1,500 1,550
		Batt. 6.	1.1	....	......	90 135 157.5	4.5 9. 10.5	3 4.5 6	9,700 8,000 7,400	7.5 7.5 7.5	775 940 1,020
UX-120 or CX-220	....	3.3	.125	....	......	90 135	16.5 22.5	3.2 7.0	7,700 6,600	3.3	428 500
UX-112—CX-112	....	5.	.5	....	......	90 135 157	6.0 9.0 10.5	2.4 5.8 7.9	8,800 5,500 4,800	8.5	890 1,435 1,670
UX-171—CX-371	....	5.	.5	....	......	90 135 180	16. 27. 40.5	11. 16. 20	2,500 2,200 2,100	3.0	1,200 1,360 1,430
DeForest DV-7	....	5.	.5	....	......	135	9.	6.	6,500	7.	....
Western Electric 205D	....	4.4	1.6	....	......	350	30	30	4,000	7.	....
DL-14	....	5.	.5	....	......	180	40.5	15	2,600	3.	1,100
DL-9	....	7.5	1.6	....	......	350 500	30 51	25 35	6,000	6.5	1,083

CHART OF TUBE CHARACTERISTICS (Fig. 153)—*Continued*

HIGH-MU TUBES

UX-240 or CX-340	Amp.	5.	.25	....	....	90 180	1. 3.	.3 .7	60,000	30	900
Ceco G	Amp.	5.	.25	....	....	90 180	1. 4.5	1. 1.8	13,600	20	1,420
Daven MU-20	Amp.	6.	.25	....	....	90 180	.5 1.5	.35 1.1	40,000	20	500
Daven MU-6	Amp.	6.	.5	....	....	90 135 180	4.5 9.–10.5 16.5–18	10.–12.	....	5–6.	1,800
Sonatron MU-20	Amp.	5.	.25	....	....	90 135	.5 1.5	.5 1.5	24,000	20	835
Sonatron MU-30	Amp.	5.	.25	....	....	90 135	.5 1.5	.3 .5	35,000	30	900
Schickerling MU-20	Amp.	5.	.25	....	....	90 200	.5 3.	.4 .9	40,000	20	500

A. C. TUBES

Sovereign A. C.	Det. Amp.	2¾–3.	1.1	22–45	.....	90 135	4.5 9.	....	....	....	....
Van Horn A. C.	Det. Amp.	.8	2.5	....	....	....	....	....	....	....	....
UX-226 or CX-326	Amp.	1.5	1.05	....	....	90 135 180	6 12 13.5	3.7 3 7.5	9,400 10,000 7,000	8.2 8.2 8.2	875 820 1,170
UX-327 or C-327	Det. Amp.	2.5	1.75	....	....	90 135 180	6 9. 13.5	3. 5 6	11,300 10,000 9,400	8.2 8.2 8.2	725 820 870

CHART OF TUBE CHARACTERISTICS (Fig. 153)—*Continued*

	Kind	Filament Voltage	Filament Current	Use	Maximum Input Voltage	Output Current
DR	Half Wave	7.5	2.	Rectifier	550	35
UX-213 or CX-313	Full Wave	5.	2.	Rectifier	220 per plate	65
UX-216B or CX-316B	Half Wave	7.5	1.25	Rectifier	550	65
UX-280 or CX-380	Full Wave	5.	2.	Rectifier	330 per plate	125
UX-281 or CX-381	Half Wave	7.5	1.25	Rectifier	750	110

MISCELLANEOUS TUBES

	Use	Voltage Drop	Current	
UX-874 or CX-374	Output Voltage Regulator	90.	50 ma.	Starting Voltage 125
UX-876 or CX-376	Ballast Tube	40 to 60	1.7 amp.	....
UX-886 or CX-386	Ballast Tube	40 to 60	2.05 amp.	....
UV-877 or C-377	Protective Tube	....	....	....

frequency or neutrodyne receivers, the best tube equipment should include a sensitive alkali-vapor detector tube of the UX 200A or CX 300A type, and a power amplifier to feed the loudspeaker, either the UX 171 or CX 371, although it will be found that the UX 112 or CX 112 are more adaptable for power amplifiers in certain circuits. The receiver will then meet the latest standards for excellence in broadcast reception.

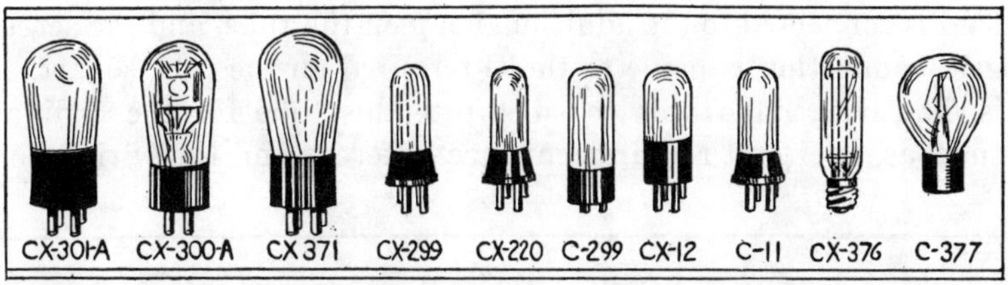

Fig. 154. A Number of Modern Radio Tubes Which Are Used for Various Purposes Such as Detection, Amplification, Power Amplification in Receivers, and Stabilization of Eliminators

The UX 120 or CX 220 tubes are power tubes used on 3-volt circuits in conjunction with UX 199 or CX 299 and UV 199 or C 299 tubes in the other stages. The WD 11 and WD 12, or the C 11 and CX 12, are also general-purpose tubes for dry-cell circuits, and they operate on a potential of 1½ volts, one or more dry cells being connected in parallel to obtain this voltage. These tubes are, however, not as efficient as the 3-volt tubes, and have no power tube to supplement them.

The UX 210 or the CX 310 tube is a power tube used to operate heavy-duty loudspeakers. The UX 213 or CX 313, and the UX 216B or CX 316B, are rectifier tubes used in socket-power devices; the 13 type is a half-wave rectifier, while the 16B type is a full-wave rectifier. The UX 874 or CX 374 tubes are called "glow" tubes and are used to regulate the output voltage of B-eliminators. The UX 876 or CX 376 tubes are "ballast" lamps, and are connected in series with the primary of the power transformer in B-eliminators. These tubes tend to keep the output voltage constant despite variations in line voltage. The UV 877 or C 377 are protective tubes, preventing the tubes in a radio receiver from burning out when short-circuited by a high voltage.

Screen-Grid R. F. Tube (UX 222 or CX 322).—A new type of radio vacuum tube, especially designed for radio-frequency amplification, and known as the UX 222 or CX 322, is now available. Its amplification factor is many times greater than that of the ordinary r.f. tube, and internal shielding by means of a screen grid makes neutralization of plate-to-grid capacity unnecessary. This tube has a standard base with four tips, and a fifth connection is provided at the top through a metal cap. There are two grids besides the usual tube elements, the additional grid being used as a shield to make the tube more stable. The usual grid lead is connected to the additional cap on the tube, and the screen-grid connection is made to the G-terminal on the tube socket. It is, however, impossible to substitute this tube for the ordinary tube, as its circuit requirements are different and the wiring must

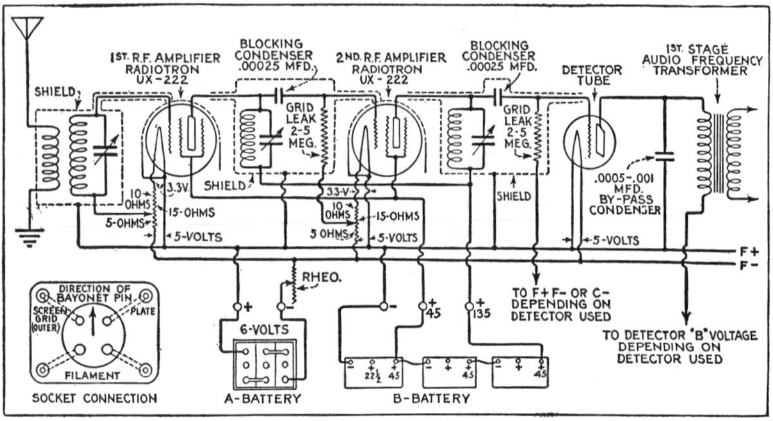

Fig. 154*a*. Diagram of R. F. Amplifier Designed for the Use of the Screen-Grid Tubes. Note the Complete Shielding Indicated by the Dotted Lines, Special Tube Shields Being Used Over the Tubes, Besides Shields for the Coupling Units

be altered to some extent. Besides, it is absolutely necessary to shield this tube effectively and also the grid leads connected to it. A filament voltage of 3.3 volts is used, and the recommended plate voltage is 135 volts. Fig. 154*a* shows the wiring diagram of a radio-frequency amplifier having two stages of r.f., using screen-grid tubes. The dotted lines indicate shielding, and the position and values of the condensers and resistors are clearly shown. The use of this screen-grid tube eliminates howling caused by uncontrollable oscillation in the r.f. stages. Most old receivers can be rewired to use the new tubes, which will increase the range and pick-up quality of the receiver considerably.

Power Tubes.—Many radio owners have the mistaken idea that the purpose of power tubes is primarily to increase the volume of a receiver. They are really designed to provide a greater amplification range than ordinary tubes, and this enables them to reproduce reception with very little distortion. The object, then, of these tubes is more uniform reproduction instead of greater volume. It should also be remembered that the addition of a power tube to a receiver does not compensate for losses due to various causes. Usually higher B-battery voltages are required for these tubes and corresponding high grid voltages. The Daven MU 6 and some UX and CX tubes, as explained in the preceding paragraph, are of this type, and they are used in the last stage of audio amplification. Others are given in Fig. 153.

Common Tube Troubles

Besides the common trouble of tubes weakened by overloading them, as already explained, there are other acute troubles, which may prevent tubes from functioning properly or may even endanger other parts of the receiver. The most common of these troubles are grid-to-filament and grid-to-plate short circuits, loose tube bases causing short circuits inside of the base, faulty contacts of socket prongs and tube tips, the incorrect use of tubes, inadequate filament supply, and "microphonics," a trouble to which many tubes are subject.

Grid-to-Filament Short Circuits.—Grid-to-filament short circuits are usually caused by the filament sagging against the grid, as shown in detail 1 of Fig. 155. This trouble is especially likely to occur in receivers where the tubes are mounted at an angle, or perhaps are held horizontally. Owing to the constant heating and cooling of the filament, and partly owing to its weight, the filament may sag, finally making contact with the grid. In case of 3-volt, dry-cell tubes, which are used extensively in portable sets, the grid surrounds the filament, and their close proximity increases the danger of such short circuits. Hissing, crackling, and frying noises will then be heard, which spoils all reception. The results of this trouble may be a burned-out secondary winding of the transformer to which the grid of the offending tube is connected. In case the tube is used for radio-frequency amplification, where successive stages are coupled with air-core transformers, the transformers are usually not seriously injured. If the receiver is subjected to considerable vibration, as is the case in portable sets or those installed in automobiles, trains, or similar places, there may be trouble from occasional short circuits

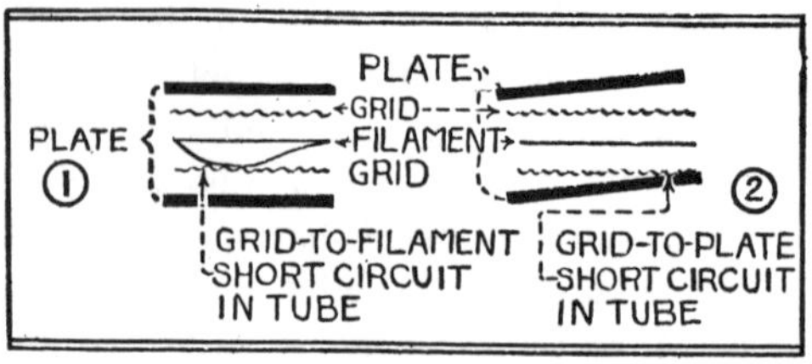

Fig. 155. (1) Grid-to-Filament Short Circuit in Horizontally Mounted Tube. (2) Grid-to-Plate Short Circuit Caused by Bent Plate

between the grid and filament, producing rasping and crackling noises. If these elements vary without touching each other, there will be fluctuations in signal intensity, corresponding to the physical vibration taking place.

Grid-to-Plate Short Circuits.—Grid-to-plate short circuits, as shown in detail 2 of Fig. 155, are fortunately not so common as grid-to-filament short circuits. When the former occur the other tubes in the receiver are in danger of being burned out as well as the windings of the transformers on each side of the defective tube. A burned-out potentiometer may also result from this trouble. The paths of the circuits in case of a defect of this kind are shown in Fig. 156. Such short circuits are possible because the B-negative line is connected to either the A-positive or the A-negative line at some point inside of the receiver or at the batteries. The grid usually carries an A-negative charge and is therefore connected to the negative side of the filament wiring. As the plate carries a B-positive potential, there would be a path for the B-current over the grid, through the winding of the transformers to which the grid and plate are connected, over one of the filament lines, and passing through one or more tubes, depending on the wiring of the receiver.

Testing for Internal Short Circuits in Tubes.—Fig. 157 shows how a pair of headphones and a 4½-volt C-battery are used to test for internal short circuits in tubes. On the later-model tubes, the heavy tips are connected to the filament, and the two lighter tips are connected to the grid and plate, as indicated in the drawing. Hold one testing tip to the grid and the other to the plate terminal of the tube. If the tube is in good condition, no click will be heard, but if clicks can be heard, there is a short circuit

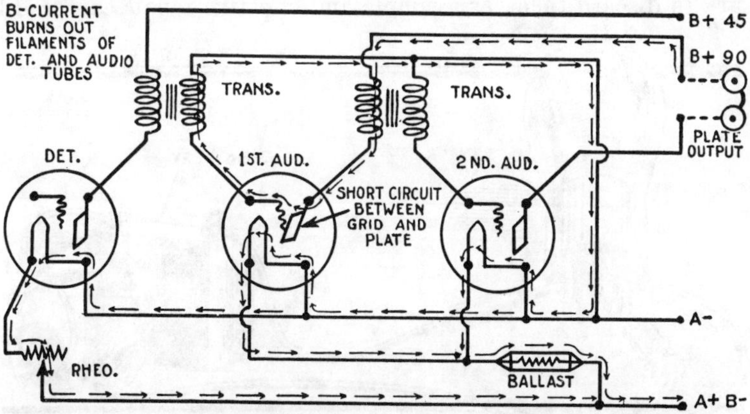

Fig. 156. Diagram Showing how a Short Circuit Between the Grid and the Plate Elements of a Tube Permits Plate Voltage to Get in the Filament Circuit and Burn out Several Tube Filaments

between the grid and the plate. Then, while still holding one testing tip on the grid, touch the other to either one of the filament terminals. A click again indicates a short circuit. Be sure not to use a testing voltage higher than the filament voltage of the tube being tested, or the filament may accidentally be burned out. Some tubes develop sporadic short circuits, which manifest themselves only now and then. To find such defects, one should gently tap the tube while testing, holding it in a horizontal position and turning it around gradually. For such tests it will be found handy to use clips on the end of the phone cords, which are snapped to the tube tips, and the operator can then manipulate the tube more easily. Tubes which have internal short circuits cannot be tested for plate-current, etc., as they may burn out the plate milliammeter if such a test is attempted. Neither can they be reactivated. They should be discarded at once and replaced with new ones.

Loose Tube Bases.—Some tubes become loose in their base, especially if they are inserted and removed from the sockets frequently. The cement holding the glass to the bakelite base sometimes dries out so that it cracks and loosens from the glass.

There is then danger of twisting the wires inside of the base when inserting or removing the tube from a socket, and such twisted wires would result in short circuits. Tubes otherwise in good condition may develop this trouble, but it is not necessary to discard them, as a simple and effective repair can readily

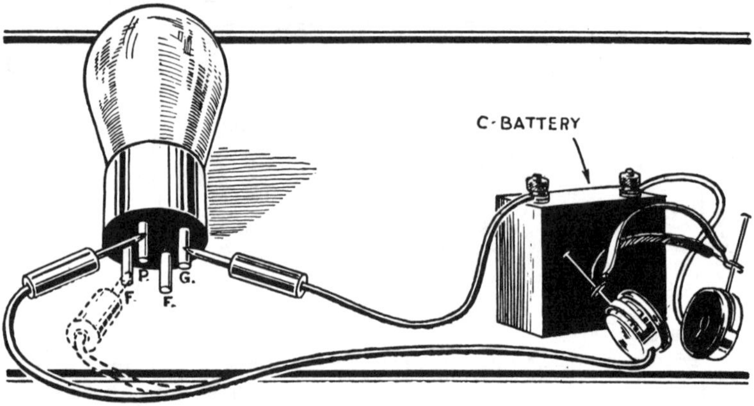

Fig. 157. Method of Using Headphones and C-Battery for Making Tests on a Vacuum Tube in Order to Determine Whether or Not It Is Short-Circuited Internally

be made. Wrap a layer of adhesive tape partly over the glass and partly over the top of the base as shown in Fig. 158. This will prevent the tube from being turned around in the base, although care should nevertheless be taken when removing it and reinserting it in the socket. Another remedy for this trouble, also shown in Fig. 158, is to scrape away the surplus cement protruding above the base, and as much as possible between the base and the tube. Then apply some collodion or "New Skin," letting it run down between the bulb and the base, and permitting a liberal film to dry on the outside over both. When collodion is so thick that it will not flow readily, use some ether to thin it down to a suitable consistency.

Faulty Contact of Tubes in Sockets.—After a receiver has been in use for some time, trouble may be experienced from faulty contacts between the tube tips and the socket prongs. Crackling and frying noises, reduced volume, and sometimes total inaudibility may result. The reason for the trouble may be the gradual formation of corrosion on the socket prongs and on the tube tips, which makes a high-resistance joint, or the trouble may be due to the socket prongs being bent down too far. The latter is often the case, especially with cheap sockets in which the prongs do not have enough "spring" and are therefore apt to bend easily. Frequent changing of tubes in sockets is also responsible for this trouble. The prongs can be lifted up

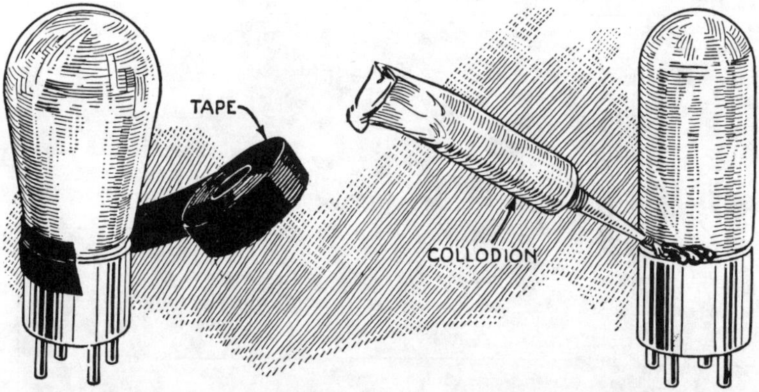

Fig. 158. When the Glass Bulb of a Vacuum Tube Becomes Loose at the Base, It Can Be Remedied by the Use of Tape or by Means of Collodion

again with the aid of an ordinary buttonhook, as shown in Fig. 159, the batteries being disconnected while this is done. Another reason for a loose socket prong is the loosening of the screw-

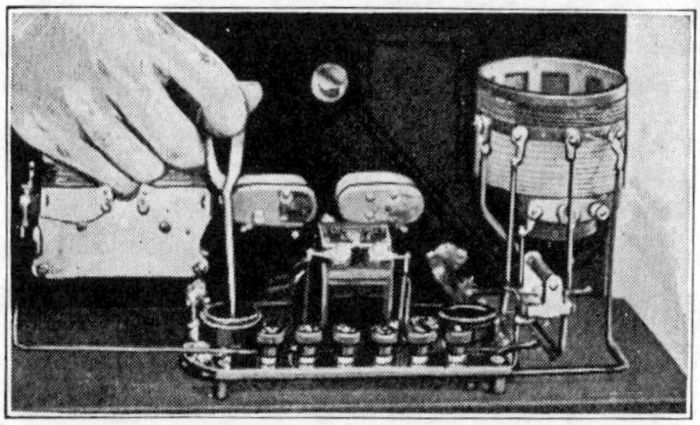

Fig. 159. A Button Hook Will often Be Found Useful for Pulling up the Prongs of Tube Sockets

(Courtesy of "Popular Mechanics Magazine")

and-nut assembly, which holds the socket prong in place. Fig. 160 shows the construction of sockets and the points where they are apt to loosen, allowing play at the prongs. Go over the sockets periodically to see that the locking nut is tight; if not, get a small socket wrench used for radio work and tighten it down securely, as shown in Fig. 160, right detail. Some types of sockets have prongs which make a "wipe" contact on the tube tips. These are sometimes so tight that the soldered ends of the tube tips are pulled off when removing a tube, and this ruins it.

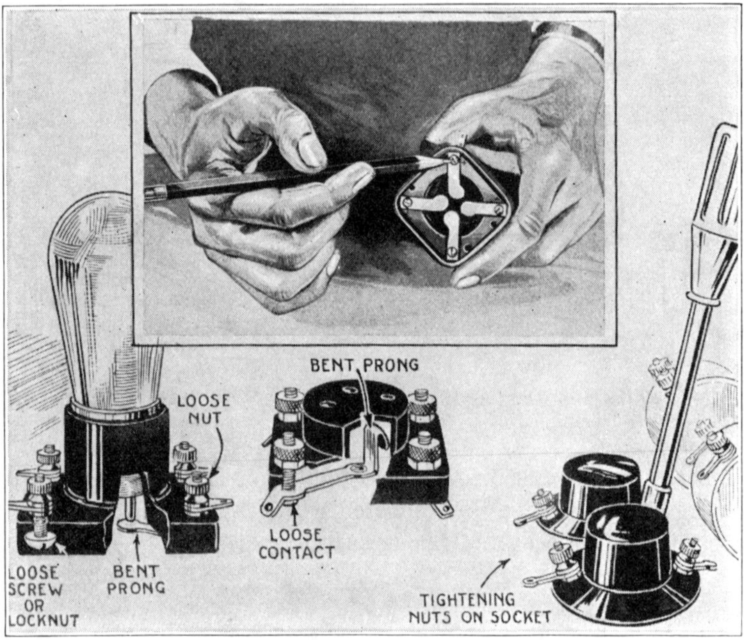

Fig. 160. When Socket Contacts Are Loose or Prongs Are Bent, Crackling Noises and Flickering Tubes May Be Observed. The Assembly of Both Old and New Types Is Shown Here

To avoid recurrence of loose nuts, get some small lock washers and insert these under the locknuts. Then replace the soldering lugs, to which the connecting wires are attached, and turn down the thumbnuts. The thumbnuts also have a tendency to get loose, sometimes as a result of vibration, especially if the receiver is being transported, and they should therefore be carefully inspected. The same procedure of using lock washers is advisable, and the thumbnuts should be replaced with hexagon nuts, these being turned down tight with a socket wrench. When the screw assembly is examined, clean off the soldering lug and the adjacent nuts with alcohol, for it often happens that a film of soldering flux will be found here, due to using an excessive amount while soldering.

When the tube tips and the socket prongs have a film of corrosion over them, the grid and plate circuits are especially apt to be affected, but the filament connections usually do not suffer from slight corrosion at these points. However, loose contacts at the filament prongs may cause the tube to flicker or go out entirely. Every radio owner should occasionally remove all the tubes from the sockets and clean off the tips with a small nail file or with a piece of emery cloth, as shown in Fig. 161. The socket

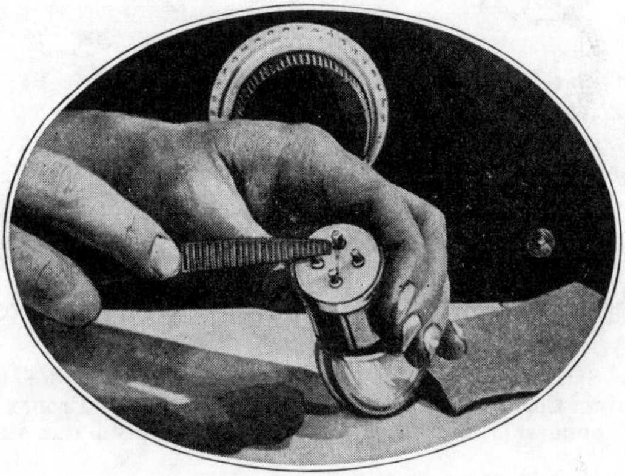

Fig. 161. Dirt and Corrosion on Tube Tips Can Easily Be Cleaned off by Means of a Small Nail File
(Courtesy of "Popular Mechanics Magazine")

prongs should be given a similar treatment, and this is best done by holding a strip of emery cloth over the end of a spudger or any flat stick about 5 or 6 in. long, which permits one to get at the prongs without any trouble. This is shown in detail 1 of Fig. 162. It is advisable to disconnect the B-battery leads while performing this work, as one of the socket prongs might be loose and by pushing it over against an adjacent prong, a short circuit might result.

Some manufactured receivers have the sockets riveted to the subpanel, as shown in detail 2 of Fig. 162. In case only one rivet is used to hold each prong in place, there is often trouble resulting from the prong becoming loose. It may then rotate on the rivet, and its end may touch an adjacent prong and cause a short circuit. Such a short circuit between the filament prongs will cause the A-battery to be discharged rapidly, and hot wires, resulting from a rapid discharge, may cause fire. A short circuit between the grid and plate or between the filament and plate prongs may result in burning out one or more tubes, and in rapid discharging of the B-battery. A burned-out secondary winding

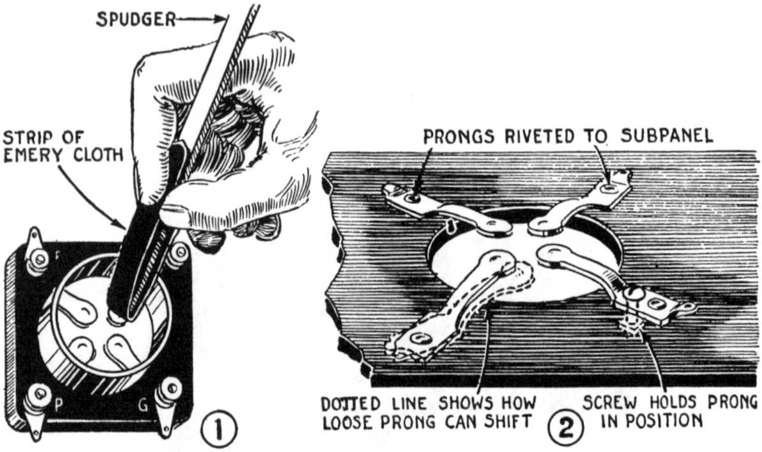

Fig. 162. (1) Strip of Emery Cloth Held over the End of the Spudger Permits Cleaning Corroded Socket Prongs. (2) Loose Prongs Riveted to Subpanel of Commercial Set Can Be Fastened Securely by an Extra Screw

of a transformer may also be caused by a grid-to-filament short circuit. To eliminate this trouble, drill an extra hole through both the loose prong and the subpanel, just behind the rivet, and use a short screw and nut to hold the prong in place securely. Do not drill the hole on the side of the rivet toward the center of the socket, because this will hold the prong much more rigidly and it is then apt to be bent when a tube is inserted.

Microphonic Action of Tubes.—Almost every radio owner who uses a "hard" tube for a detector, has, at some time or other, experienced a loud, howling noise, which increases in volume and rises in pitch, making reception an impossibility. This is what is known as a "microphonic," a noise caused by one or more vibrating tubes, the detector usually being the offending one. To make sure whether the trouble really is a microphonic, grasp the tube securely while the noise is heard; if it stops the trouble has been located. As just stated, the cause of a microphonic is vibration of the tube. There may be a number of sources of such vibration, and the removal of the source is the most positive remedy. This, however, is not always possible, as the source may be the loudspeaker, and without this one could have no reception at all. Then again the vibration may be caused by street traffic, by machinery in the building, by persons walking around, or even by a battery eliminator.

The vibration is transmitted through the walls, floors, table, or stand on which the receiver is placed, or it may even be transmitted through the air. Sometimes there seems to be no external source of vibration at all, and the tube just starts to produce the noise, seemingly of its own accord. The quality of each tube is slightly different from that of another, and one tube may therefore be much more microphonic than another, a variation of uniformity which can naturally be expected when tubes are manufactured by the thousands in rapid quantity production. Also, it will be found that the small, 3-volt, dry-cell tubes have a greater tendency toward being microphonic than the large 5-volt tubes.

Sometimes the trouble can be cured by merely "switching" tubes, or by decreasing the plate voltage of the detector or of the first audio tube. There are also many devices on the market

Fig. 163. To Eliminate Vibration Causing Microphonic Action of Tubes Sponge-Rubber Cushions Should Be Provided under the Receiver Cabinet, under the Base of the Loudspeaker, and under the Base of the Eliminator

for curing this trouble. These are usually heavy shields or weights to be placed on the tube, or some means of holding the tube rigidly in position. It is better to arrest the vibration before it enters the set. This can be done in various ways. Get a few pieces of 1-in. sponge-rubber and provide cushions of this material under both the receiver cabinet, the eliminator, and the loudspeaker, as shown in Fig. 163, the latter being the most common cause of microphonic trouble. The provision of cushion sockets, as shown in Fig. 164, instead of rigid sockets will also

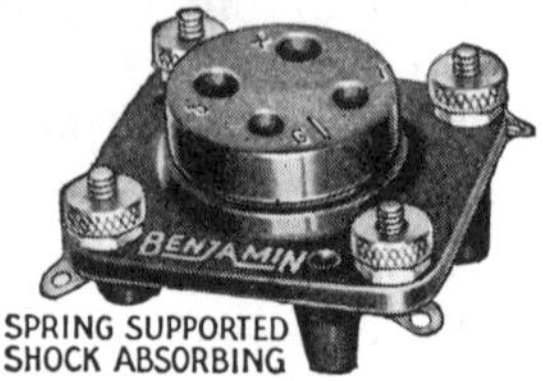

Fig. 164. Shock-Absorbing Socket Helps to Prevent Microphonics

be found an effective method of eliminating this trouble. If this is not entirely effective, remove the loudspeaker as far as possible from the set, preferably on another stand or table. Also be sure to turn the bell of the loudspeaker away from the set, instead of toward it, for it has been found that the air vibration caused by reproduction of sound by the loudspeaker is transmitted to the tubes, as shown in Fig. 165, and in many cases this alone is enough to cause them to vibrate.

Fig. 165. Air Vibrations Caused by Loudspeaker, Especially Those of the Horn Type, Make Tubes Vibrate Slightly, Which in Turn Produces a Microphonic

These measures are certain to reduce the trouble considerably, and in most cases will eliminate it entirely. But should it still persist, change the hard detector for a detector tube of the "soft" type, such as a UX 200A or CX 300A tube. These tubes are less microphonic, will give better tone quality, and are much more sensitive than the "hard" tubes. With the popular use of consoles nowadays, much trouble from microphonics is experienced by the proximity of the built-in horn to the receiver, and its rigid attachment to the console. In such cases the speaker should be unfastened from the console, and then suspended less rigidly. One method of doing this is shown in Fig. 166.

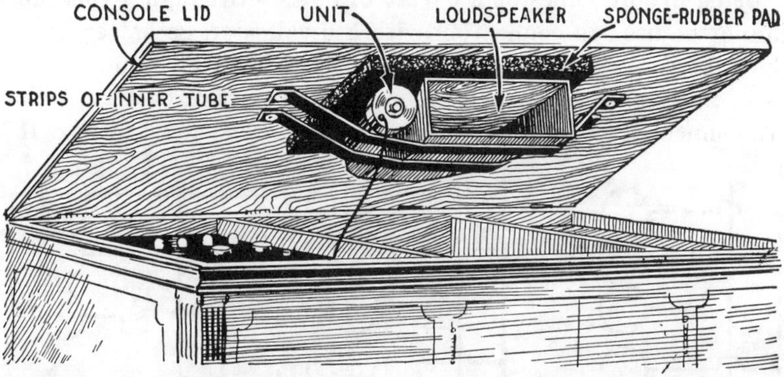

Fig. 166. Horns Attached Rigidly to Consoles often Cause Microphonics. This Trouble Can Be Remedied by Suspending the Horn in a Semi-Rigid Fashion, Using Sponge-Rubber and Pieces of Inner Tube

Using Wrong Tubes and Ballasts.—A "soft" tube of the UX 200A type will not operate as an amplifier, and power tubes usually do not make detectors. Ordinarily 3-volt tubes cannot be used on circuits intended for the use of 5-volt tubes unless changes are made in the circuit, and 5-volt tubes will not operate on circuits designed for 3-volt tubes. When renewing tubes be sure to replace them with the same kind of tubes that were originally used. If it is desired to substitute a power tube for an ordinary UX 201A or similar type of tube formerly used in the last audio stage, be sure to provide the correct C-bias, and also the correct plate voltage. Some power tubes are designed to operate from the filament lines directly without the use of a rheostat or ballast. If a rheostat is connected in the line supplying the power tube it can be turned up entirely, provided no other tubes are controlled by it, but in case a "ballast" or fixed resistance is used in the line supplying the last tube, this must be removed and the wires formerly connected to the ballast must be connected together in order to get the full filament current for the power tube. Whenever tubes of different filament-voltage requirements are substituted for the original tubes of a receiver, be sure to provide ballasts of the correct sizes to supply the tubes with the required voltage. Too much or too little voltage will not only prevent the tube from operating efficiently, but may also injure the filament. It should also be remembered that ½ and ¼-amp. tubes cannot be connected to one ballast.

Using 3-Volt Tubes on a 6-Volt Circuit.—Although 3-volt tubes cannot be used in connection with a 6-volt storage battery when the tube sockets are hooked up in parallel, which is usually the case, they can be operated from a 6-volt storage battery if they are connected in series, or in series-parallel as shown in Fig. 167,

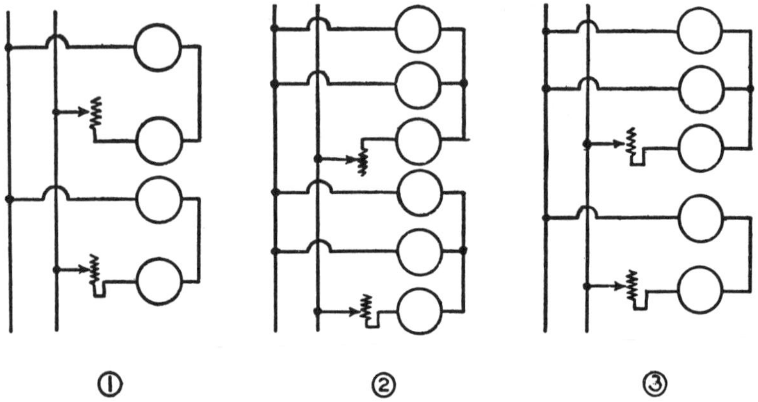

Fig. 167. Using 3-Volt Tubes on a 6-Volt Supply: (1) Two Pairs of 3-Volt Tubes Connected in Series. (2) Two Sets of Three 3-Volt Tubes Connected in Series-Parallel. (3) Five 3-Volt Tubes

the exact hookup depending on the number of tubes used. If it is desired to hook up four 3-volt tubes on a 6-volt circuit, connect the tubes in pairs as shown in detail 1. Six tubes are connected up in sets of three, as shown in detail 2. Usually 3-volt tubes will not work satisfactorily in receivers designed for the use of 5-volt tubes, because the amplification value of the former is much less than that of the latter. For this reason a considerable decrease of volume will be noticed when the small tubes are substituted for the larger ones. However, the small tubes draw a very slight current, and if a storage battery is not desired, they can be operated on ordinary dry cells. It is claimed that the use of the small tubes in radio-frequency stages of a receiver, causes less trouble from uncontrollable oscillations, manifested by whistles and howls, than the use of large tubes, for the latter have a much greater internal capacity, which is the source of this trouble. If it is desired to use small tubes in a receiver designed for large tubes, it will of course be necessary to make a few changes in some of the current and voltage-controlling devices, such as rheostats and ballasts, and if a 6-volt battery is to be used, changes must also be made in the wiring. Where 6-ohm rheostats are used for large tubes, substitute 25 or 30-ohm rheostats when the small tubes are used. Ballasts of .06-ampere capacity are used in series with the 3-volt tubes, instead of .25-ampere ballasts, which are intended for use with the 5-volt tubes of the UX 201A type.

A.C. Tubes Eliminate A-Battery

To dispense with the A-battery entirely and also to eliminate socket-power devices for providing filament current, a tube is on the market, which operates from 110-volt alternating current, stepped down to a suitable voltage for the tube by means of a small transformer. Fig. 168 shows a tube of this type and its

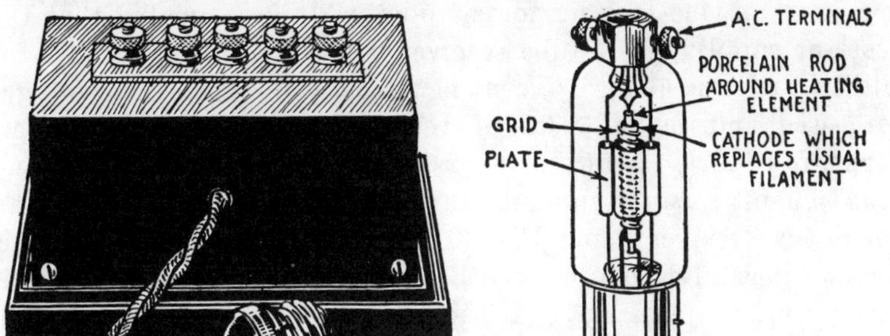

Fig. 168. A. C. Tube in Which a Cathode Replaces the Usual Filament, the Cathode Being Heated Indirectly by Means of an A. C. Filament, Which Receives Current from a Small Step-Down Transformer Shown at the Left

internal construction. It has the grid and plate elements like those of a standard tube, but it has no similar filament. Electronic emission is obtained from a plate or cathode, which replaces the usual filament, it being heated to throw off electrons. It is heated indirectly by means of a special filament which is lighted by the alternating current supplied by the transformer. The filament is, however, separated from the cathode by a porcelain tube. This arrangement permits the filament to heat up the cathode without coming in actual contact with it, and the heated cathode can then pass a plate current. The tube base fits any standard radio-tube socket, but actual electrical contact is made only with the negative filament line of the receiver, and the positive line may be removed. The negative B-supply is connected with the negative filament line. The positive B-potential is applied to the plate of the tube in the usual way.

The transformer used with these tubes, which is also shown in Fig. 168, is provided with taps for obtaining various voltages from $3\frac{1}{2}$ to 5 volts. In a receiver having from one to three tubes, the 4-volt tap is used, and in receivers having from four to six tubes, the 5-volt tap is used. The alternating current is supplied to the tube filament through the top of the tube, two terminals being provided here for this purpose. As there is no polarity on alternating-current lines, it does not make any difference to which terminals the lines are connected. The tubes are hooked up in parallel in respect to the alternating-current supply.

No rheostats are used with these tubes, a feature which appeals to anyone looking for simplicity. It is estimated that the cost of supplying a 5-tube receiver with alternating current for its operation is about 1/5 cent per hour, when normal B-voltages are used, and the audio stages are transformer coupled. An average life of 1,200 operating hours is claimed for these tubes, and the amount of plate-current draw is much more than that of an ordinary receiver using UX 201A or CX 301A tubes. For this reason it is advisable to use a B-eliminator instead of B-batteries for providing plate current. Fig. 169 shows a 5-tube tuned radio-frequency receiver using a.c. tubes.

The disadvantage of these tubes is that an alternating-current hum is often present, its degree of intensity depending to some extent on the design of the receiver, in which the tubes are used. This trouble, it is claimed, can be overcome by using a.c. tubes having filaments that do not come to full incandescence and consume a very small amount of current. Although the volume obtained from a.c. tubes is greater than that obtained from UX 201A or CX 301A tubes, the distance-getting qualities are, as a rule, perceptibly less, owing to the hum characteristic. For the radio owner who is satisfied with local reception and does not seriously resent a slight hum, a receiver employing a.c. tubes will be entirely satisfactory.

If it is desired to adapt a receiver equipped with standard tubes to the use of a.c. tubes, all that is necessary is to connect the B-negative to the A-negative line in the circuit, if it is not already connected in this way. In case the B-negative line was previously connected to the A-positive line, disconnect the two.

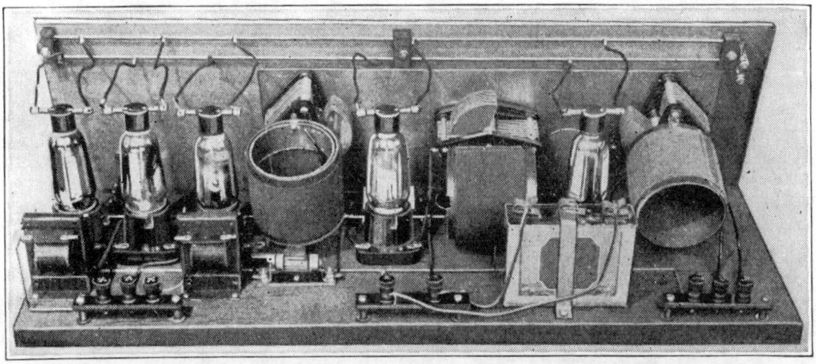

Fig. 169. A 5-Tube R. F. Receiver Using A. C. Tubes, Which Eliminate the Necessity of an A-Battery. The Current Supplying the Tubes Is Carried in Bus-Bar Conductors Attached to the Panel

(Courtesy of "Popular Mechanics Magazine")

The entire A-positive line can also be removed if this is desired. Then install the supply lines from the step-down transformer. It will be found convenient to use No. 14 bus-bar wire, ordinarily used in radio-receiver construction, for this purpose, attaching it to the panel or subpanel of the receiver. Do not get these lines close to the coils or to the grid and plate leads, as this may cause considerable trouble due to induction. Solder ordinary lugs to short lengths of stranded, insulated wire, and solder the other end of these lengths of wire to the supply lines.

Three-In-One Tubes

Special tubes having three separate filaments, grids, and plates, thus really being equal to three tubes in one, have been put on the market. A tube of this type, shown in Fig. 170, draws .25-ampere of filament current, and has a thoriated filament permitting it to be reactivated. It operates on a potential of 5 volts for the filament and from 45 to 135 volts for the plate. The base is standard, so that it fits in any standard radio-tube socket. Four binding posts are provided, permitting connection to be made to transformers or other points in the circuit. Such a tube can be used in any number of circuits and hookups, as, for instance, a receiver having one stage of radio-frequency amplification, a detector, and two stages of audio-frequency amplification, the first stage of audio-frequency being reflexed on the radio-frequency

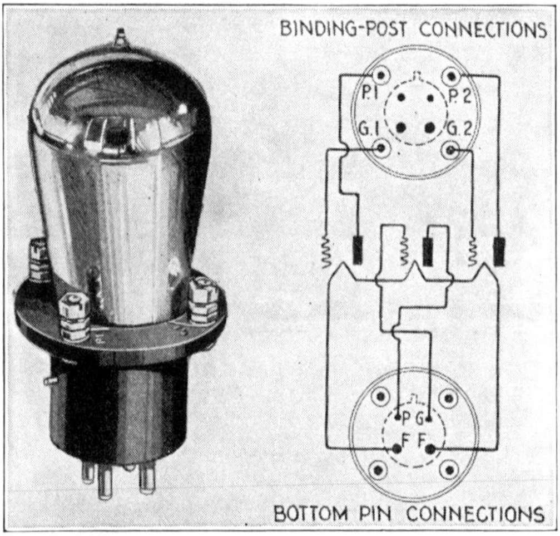

Fig. 170. Multi-Valve Tube Having Three Sets of Filaments, Grids, and Plates, Which It Is Claimed, Give It the Efficiency of Three Separate Tubes

(Courtesy of "Popular Mechanics Magazine")

stage. This makes a receiver capable of giving good distance reception and loudspeaker volume. Complete instructions for its use are provided with the tube when sold.

Multiple tubes, somewhat similar to the one described in the previous paragraph, but having two stages of resistance coupling also incorporated, as shown in Fig. 171, are being manufactured in Germany. The filament potential is 4 volts and the filament current is 3 amperes. No rheostat is needed. The plate voltage is from 90 to 150 volts, and at 90 volts plate potential the plate current is from 3 to 5 milliamperes. One tube of this type in a receiver, it is claimed, will give loudspeaker volume. The base has six contacts and requires the use of a special socket. In size it is slightly larger than power tubes of standard make.

Bismuth-Plate Detector and Amplifier Eliminates Tubes

Complete elimination of tubes and batteries is the claim made for a small device shown in Fig. 172. It consists of a number of

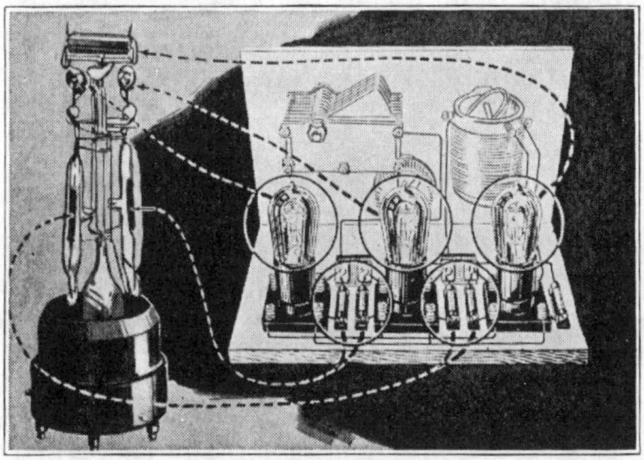

Fig. 171. Multi-Valve Tube of Foreign Make Having Two Stages of Resistance-Coupled Amplification Besides Three Sets of Tube Elements

(Courtesy of "Popular Mechanics Magazine")

Fig. 172. Bismuth-Plate Detector Which Eliminates All Need for Batteries and Tubes in Radio Receiver

(Courtesy of "Popular Mechanics Magazine")

bismuth plates protected from atmospheric corrosion by a coat of sulphur, and a pick-up coil of insulated wire wound around them, five wires issuing from the device, with which connections are made to the receiver on which it is used. It can be used either as a detector or as an amplifier, and requires no outside current for its operation except the radio signal. It has been found by rough laboratory experiments that the device is about four times as sensitive as an 8-tube receiver. At the time of this writing, however, the device is not yet on the market and is therefore not available to the public.

General Care of Tubes

Always treat the tubes with care; never tap them severely or put them down with a jar. Dropping a tube nearly always ruins it, even if the glass does not break and even if the filament still lights. Keep the ends of the tube tips clean with fine sandpaper or emery cloth to prevent corrosion, which sets up a high electrical resistance and reduces the efficiency of the receiver considerably. It should be remembered that tubes are the most important factor in a radio receiver and should always be kept in good condition. They should be replaced if defective, in order to obtain good reception.

CHAPTER VII

INTERNAL DISORDERS IN RADIO RECEIVERS

Like any other piece of electrical apparatus, the radio receiver itself is subject to wear and depreciation, since it is made up of a number of delicate and fragile instruments, which may become defective or which may break down entirely. Movable parts, such as variable resistance devices, tuning devices, and condensers, may become loose, permitting undesirable play, and possibly short circuits or open circuits. The wiring, too, may develop faults after a receiver has been in use for some time, especially if the wiring has been done carelessly, as shown in Fig. 173. In other parts, such as jacks and switches, where electrical

Fig. 173. Careless Wiring and Poor Soldering in Radio Receivers Is Apt to Give Rise to Considerable Trouble
(Courtesy of "Popular Mechanics Magazine")

connection is made by wire contacts, trouble may result from a film of corrosion, which acts as insulation. The particular operating conditions to which a receiver is subjected may also entail difficulties. However, a receiver should never be taken apart, that is, the chassis should not be removed from the cabinet for close examination and testing, until all the accessories have been thoroughly examined and tested, for most radio troubles are found in the accessories. When the owner is satisfied that the accessories are in good condition, the tubes should be removed from their sockets, the aerial, ground, and battery leads should be disconnected at both the receiver binding posts and the battery binding posts, and the loudspeaker should be removed from the receiver. Then the chassis can readily be slipped out of the cabinet.

Receiver Operation and Terminology

In this section parts of the receiver are referred to, and if the radio fan is not already acquainted with receiver terminology, the following paragraphs will serve to make subsequent references clear.

Radio-Frequency Amplification.—By radio-frequency (r.f.) amplification is understood the strengthening and building up of high-frequency radio currents by means of vacuum tubes and coils, transformers or resistors. Radio-frequency amplification precedes detection, the latter changing the radio-frequency impulses to trains of audio-frequency (a.f.) impulses. These have

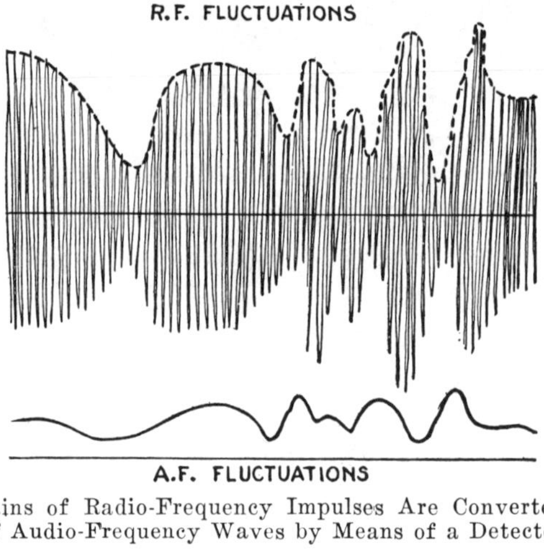

Fig. 174. Trains of Radio-Frequency Impulses Are Converted to a Series of Audio-Frequency Waves by Means of a Detector

a comparatively slow rate of fluctuation, and they are converted to sound waves by the headphones. The transformation from radio-frequency vibrations to trains of audio-frequency vibrations is illustrated in Fig. 174.

Tuned Radio-Frequency Amplification.—Tuned radio-frequency (t.r.f.) receivers are those having one or more stages of radio-frequency amplification, using air-core coils as coupling between successive tubes, and variable condensers connected across the secondary windings of these coils to tune them in resonance with incoming signals, as shown in Fig. 175. Sometimes the coils and condensers are combined to form one unit. Receivers of this type are extremely popular today. Usually they have two stages of

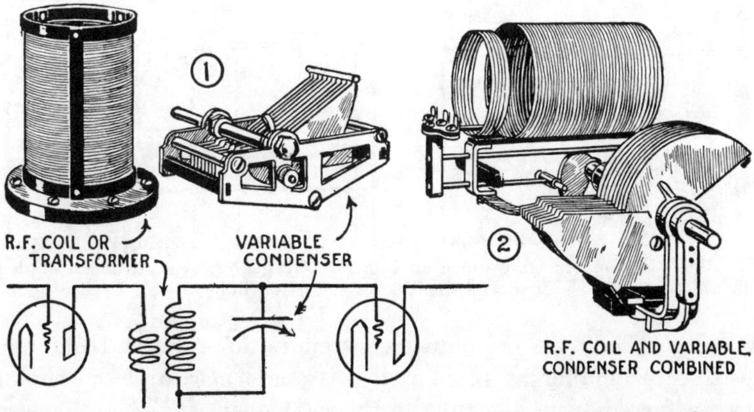

Fig. 175. Coils and Condensers Used in Tuned Radio-Frequency Stages: (1) Separate Coil and Condenser. (2) Combined Coil and Condenser. The Diagram Shows How the Units Are Wired

radio-frequency amplification, which requires the use of three coils and condensers. However, there are many receivers that have the condensers in tandem, or on one shaft, so that they can all be operated from one dial on the panel.

Audio-Frequency Amplification.—This is the magnification of audio-frequency impulses by means of vacuum tubes, that are connected by means of suitable transformers, resistors, or impedance devices.

Transformer-Coupled A. F. Amplification.—This term indicates one or more stages of audio-frequency amplification, having iron-core transformers between two successive tubes, the plate output of one tube being connected to the primary winding of the transformer, which in turn is connected to the B-positive line to provide the plate with a high positive voltage. The secondary winding of the transformer is connected to the grid of the next tube and to a line carrying a negative grid-biasing voltage, usually taken from the A-negative filament line. Two stages of transformer-coupled audio amplification are shown in Fig. 176. The coils or transformers between radio-frequency tubes are connected in the same way.

Resistance-Coupled A. F. Amplification.—This method of amplification makes use of resistors of various values between the audio tubes, each resistor being substituted for a transformer

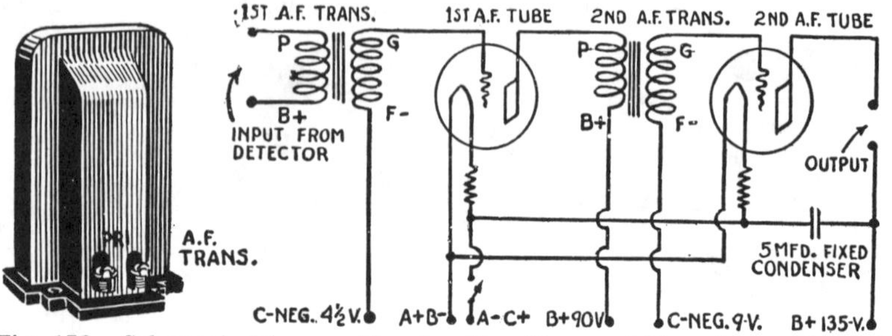

Fig. 176. Schematic Diagram Showing Two Stages of Transformer-Coupled Amplification, Designed for the Use of a Power Tube in the Last Stage

winding. There is no inductive step-up of voltage from one resistor to another as there is in a transformer, and the signal is transferred from one tube to the next over a fixed condenser. Such a method of coupling is shown in Fig. 177. The amplification of one stage of resistance coupling is less than that of one stage of transformer-coupled amplification, and three stages of the former are about equal to two stages of the latter.

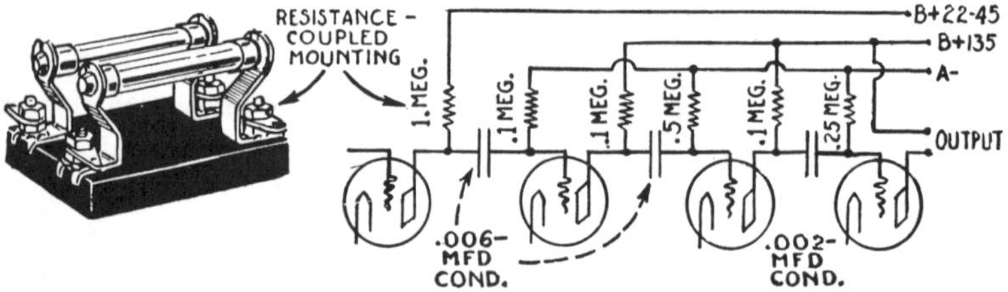

Fig. 177. Three Stages of Resistance-Coupled Amplification Are Approximately Equal to Two Stages of Transformer-Coupled Amplification. Each Coupling Unit Consists of Two Resistors and a Small Fixed Condenser

Impedance-Coupled A. F. Amplification. — Impedance devices are used instead of transformers or resistors in some circuits. They consist of one or more choke coils to suppress undesired frequencies. The output of three stages of impedence-coupled amplification is about equal to that of two stages of transformer-coupled amplification.

Preliminary Testing of the Receiver

Trouble in receivers can often be localized by making a number of preliminary tests. Fig. 178 shows a number of places in receivers where trouble may develop. Transformers and coils can be tested for continuity by means of a headset and a small C-battery, connected in series. A click should be obtained when the tester tips are held across the terminals of a tuning coil, across the ends of the primary winding or secondary winding of radio-frequency and audio-frequency transformers. The absence of a click indicates an open circuit, which may be the seat of the trouble. All resistance devices, whether fixed or variable, should show continuity of the resistance element. There should be no click when the testing tips are applied across the primary and secondary windings of a coil or transformer, as shown in Fig. 207.

Condensers can be tested by touching the testing tips to the terminals. Usually there will be a click, resulting from the current rush, but there should be no more clicks when the testing tips are repeatedly touched to the same terminals. Both fixed and variable condensers are tested in the same way. Care must be taken to see that there is not a shunt circuit around the condenser through which the current can pass, which would result in a faulty test.

The next step in preliminary testing is to look over the wiring connections to see if any of the terminal nuts are loose or if any two wires previously connected have come apart, which sometimes results from poorly soldered joints. The most frequent disorders occurring in receivers are open-circuited primary windings or audio transformers, short circuits between the plates of variable condensers, faulty resistors, and loose connections. If the preliminary test gives a clue to the trouble, it can be remedied by following the instructions contained in this section. If, however, the preliminary test does not disclose the trouble, the individual parts of the receiver must be examined separately. A good method of testing radio-frequency stages is to tune in a signal on the receiver. Then put the hand on each coupling coil; if the signal decreases in volume, the particular stage under test is in good condition, but if the volume is not decreased, this stage is not operating properly.

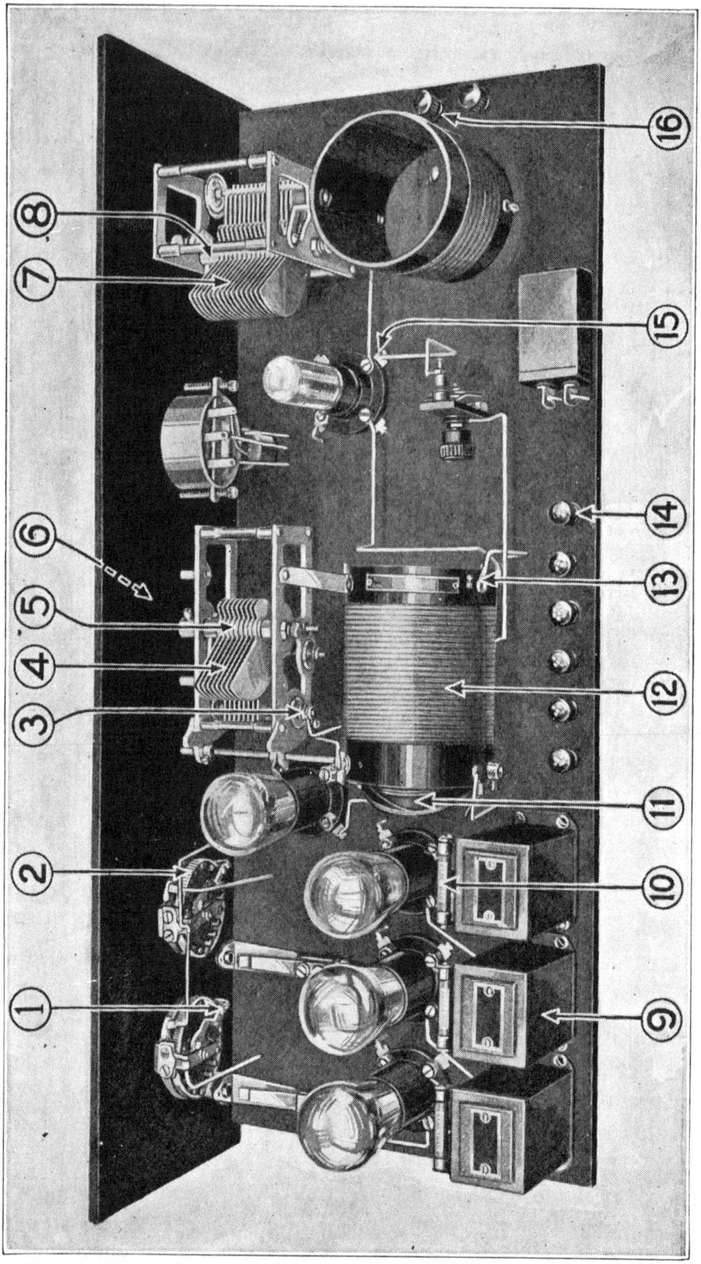

Fig. 178. Some of the Places Where Trouble May Occur in Receivers: (1) Rheostat Slider not Making Proper Contact, (2) Rheostat Resistance Burned out or Otherwise Open-Circuited, (3) Loose Nut, (4) Condenser Plates Short-Circuited, (5) Play in Condenser Shaft, (6) Dial Scrapes or Slips, (7) Rotor of Condenser Connected to Grid, (8) Broken Connection from Rotor to Terminal on Frame, (9) Open-Circuited Transformer, (10) Defective Resistor, (11) Tickler Connections Reversed on R. F. Coil, (12) Open-Circuited Secondary of R. F. Coil, (13) Improperly Soldered Joint, (14) Incorrect B-Voltage, (15) Bent Socket or Loose Assembly, (16) Ground or Aerial Disconnected, Reversed or Broken

Filament-Circuit Troubles

The filament circuit of a receiver is the network of wires that supply the current for lighting the tubes and provide grid bias, as shown in Fig. 179. Naturally all the instruments that are connected in series with these lines, such as switches, jacks, rheostats, and ballasts, are also included in the filament circuit. When the

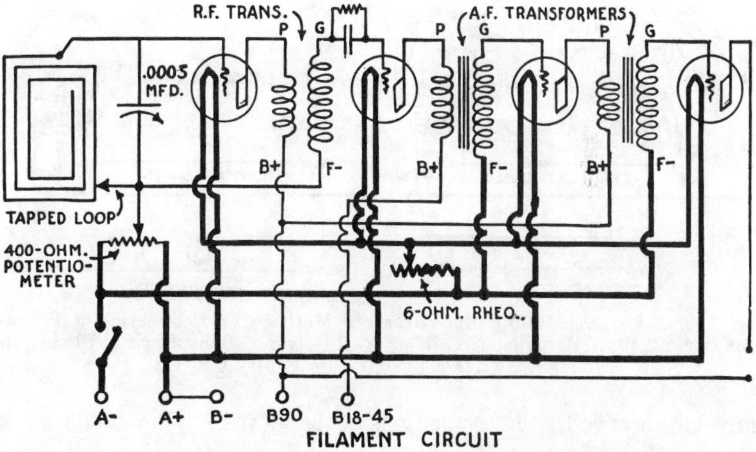

Fig. 179. Diagram of a Typical 4-Tube Loop Receiver, the Filament Circuit Being Indicated by the Heavy Lines

A-battery or A-eliminator is functioning correctly, supplying the necessary filament current at the binding posts on the receiver, or at the cable ends, and the tubes nevertheless fail to light, an open-circuited condition is evident, which can readily be detected by the headset tester; clicks indicate continuity of a conductor between the points where the testing tips are applied. As soon as there is an absence of click between points that should be connected, the trouble has been found. The places in which to look for open circuits are switches, jacks, rheostats and ballasts. Apply the testing tips to both terminals of these devices when they are in position. A click should be heard, and the absence of a click indicates that there is a break or open circuit in the device. Short circuits in the filament lines will quickly run down the A-battery. If a receiver is wired with bare bus-bar wire, it is advisable to use "spaghetti" insulation over the wire at points where the wires run close together or close to instruments.

Filament-Circuit Troubles Due to Switches and Jacks.—An open circuit may exist owing to a faulty filament switch, as shown in Fig. 180, the spring blade not making contact with the plunger. This defect prevents any of the tubes from lighting. If the blade of the switch makes a loose contact with the plunger, crackling noises will be heard in the loudspeaker, and flickering of the tubes

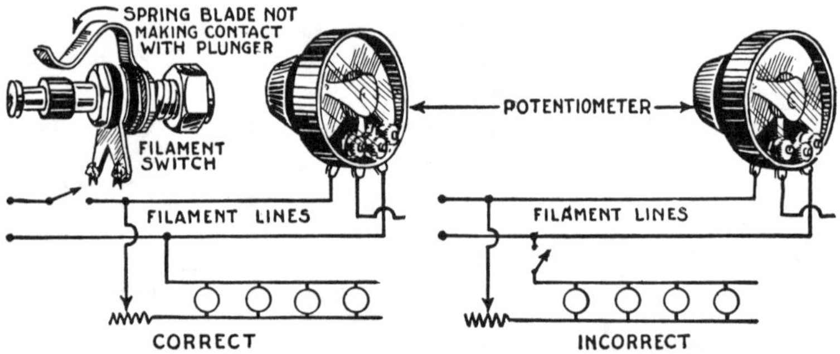

Fig. 180. The Correct and the Incorrect Methods of Connecting a Potentiometer in the Filament Circuit of a Receiver. Bent Spring Blades in Filament Switches often Cause Faulty Contact

may be observed. The remedy is to bend the spring blade back into position, so that it will make a firm connection with the plunger. If this is impossible, substitute a new switch. When replacing an instrument, be sure to disconnect the batteries from the receiver and remove the tubes in order to prevent accidental short circuits. Filament-control jacks are merely switches; when the loudspeaker plug is inserted the contacts on two spring blades are pushed together and this closes the circuit, one of the blades being connected to one of the filament-supply lines, and the other blade being connected to a line that runs to the tubes.

Potentiometer Losses.—In many receivers a potentiometer is used to vary the polarity and voltage applied to a grid. This device resembles a rheostat except that it has three terminals and has much smaller resistance wire. The two outer terminals are connected to the A-negative and A-positive sides of the filament wiring and the slider terminal is connected to a grid circuit. Some potentiometers have two sliders to obtain various voltages for supplying grid bias of different values to various tubes. If a potentiometer remains connected across the lines of the filament circuit when the receiver is not in use, there may be a slight current flow through it from the positive toward the negative terminals, the exact amount of current flow depending on the value of the resistance of the potentiometer. Although slight, such a loss is a continual drain on the battery, and it is advisable to connect it in series with the filament switch. Fig. 180 shows both the correct and the incorrect method of connecting a potentiometer in a receiver circuit.

Rheostat Troubles.—A rheostat consists of a resistance wire wound on a circular form of insulated material, one end of the winding being connected to one terminal, and the other terminal to a slider, which makes contact with the resistance wire, and can be rotated to obtain any resistance value desired within the range of the rheostat.

Open circuits sometimes occur in rheostats. This trouble can easily be detected by means of a continuity test with the headset

Fig. 181. Continuity Tests on Rheostats and Other Resistors Can Be Made with a Headphone Tester and a C-Battery

and C-battery, the testing tips being held to the terminals as shown in Fig. 181. Rheostats may be open-circuited owing to a loose slider, as indicated in Fig. 182. Sometimes the shaft of the rheostat may have developed some play, or the whole assembly may have become loose on account of handling or constant vibration. The slider may also have been accidentally bent, which usually can readily be remedied by bending it back to its normal position.

In some cases, after a receiver has been in use for some time, the resistance wire may become coated with an oxide, caused by the heating of the wire. The first symptom of this trouble is a scratchy sound when the slider is rotated over the wire. In extreme cases, parts of the winding may be covered with so much oxide that no current will flow from this section of the winding to the contacting slider. To prevent this trouble, take a rag or small brush, moistened with very light oil, and rub off the oxide from the winding along the contact path, being careful not to bend the slider. Wipe off the oil afterward with a cloth moistened with alcohol.

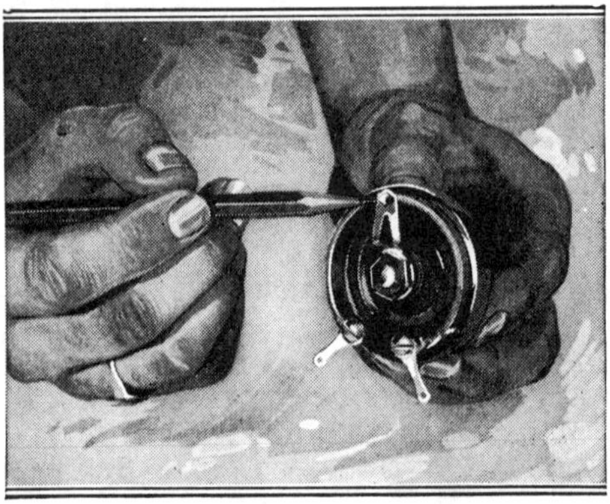

Fig. 182. Loose Sliders on Rheostats and Other Variable Resistors May Cause Crackling Noises

(Courtesy of "Popular Mechanics Magazine")

Burned-out windings of rheostats may result if too much current is drawn through them. This may occur when a UX 201A or CX 302A tube is placed in a socket of a receiver ordinarily using UX 199 or CX 299 tubes. The 25 or 30-ohm rheostat used with the small tubes cannot pass the amount of current required by the large tubes, and may consequently burn out. When replacing a rheostat, be sure to get one of the same resistance value and also of the same make or type if possible, for it can then be installed in the receiver without making any changes in the wiring. UX 201A and similar tubes require 6-ohm rheostats, while UX 199 and similar 3-volt tubes require 25 and 30-ohm rheostats. When more than one tube is connected to a rheostat, as is often the case, a lower resistance value is sometimes necessary. Do not attempt to judge the proper resistance value, but get one of the same value as the defective one to be replaced.

Defective Ballasts.—A fixed-resistance device, consisting of a resistance wire, contained in a small glass tube or capsule, and held in the receiver by means of spring clips attached to a special mounting, as shown in Fig. 183, is very popular as a substitute for rheostats. It is a step toward the elimination of variable controls on receivers, which means simplicity and stability of operation. Besides this, the use of ballasts in filament lines of receivers prevents overloading tubes by too much filament voltage. Ballasts can be obtained in various resistance values for use with one or more tubes of all types.

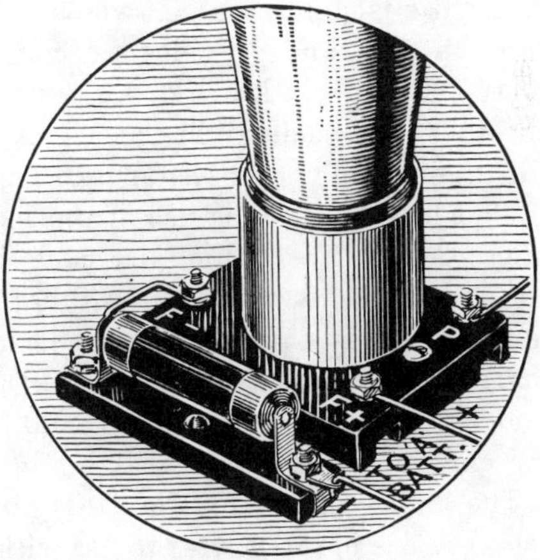

Fig. 183. Fixed Resistors of Determined Values Can Be Used on Most Stable Tubes

Sometimes the resistance wire inside of a ballast touches the glass, which may then be cracked by the heat, or the label on the ballast may be charred or burned off, which entails the possibility of fire. For this reason the size and type of ballast, which is indicated on the label, should be scratched on the bakelite subpanel, and the label then removed to eliminate this danger. The connection between the resistance element and the caps sometimes becomes loose and causes crackling noises, and this condition may also cause the tubes to flicker. If the resistance element breaks entirely, there is an open circuit, and the tubes that obtain their current supply through the ballast will not light. Always be sure that the spring clips hold the ballast securely. The clips can readily be bent toward each other so that they will grip the ballast tightly.

Using the correct value of ballasts in a receiver is just as important as using the correct value of rheostats. Ballasts are designed to carry definite currents and are to be used for certain tubes or combinations of tubes. For instance a ¼-amp. ballast is used with one UX 201A or similar type of tube, a ½-amp. ballast with two tubes, and a ¾-amp. ballast with three tubes of this type. A UX 201A or other similar type of 5-volt tube cannot be used with a ballast intended for use with a UX 199 tube.

It is true, taking into consideration the slight differences in tube characteristics, that all tubes do not operate at their full efficiency on a fixed resistance value, which allows a definite amount of current to pass, and provides a certain voltage. Some tubes operate better at slightly lower or higher current and voltage values than other, and such slight variations can not usually be obtained with the use of fixed resistances. This objection has been overcome, it is claimed, by the use of thermo-electric resistance devices such as Amperites, which are made in various sizes, according to the requirements of different tubes. The resistance element in these devices is coated with a special oxide and is enclosed in a hermetically sealed glass tube, filled with an inert gas. The element is suspended between two solid lead wires, passing through the ends of the glass cartridge and soldered to the contact caps. The table shown in Fig. 184 gives information regarding the correct size of Amperites to use with different types of tubes. Ballasts and other fixed resistance devices are more suitable for audio and radio-frequency tubes, than they are for detectors. The use of variable rheostats is generally more suitable for detector-tube operation.

Use of Fuses in Filament Circuits.—To protect tubes against being burned out by accidental short circuits, fuses should be used. Some types of fuses are connected in the filament lines at the binding posts. However, this method of fusing protects the tubes from short circuits caused by wrong connections at the batteries only, and does not protect the individual tubes from short circuits arising in the receiver itself, such as those caused by metal tools touching wires of opposite polarity. A better method is to provide a fuse for each tube. This fully protects the tubes from short circuits in the receiver as well as those caused at the batteries.

Pilot-Lamps and Dial Illumination.—Many radio owners oc-

Tubes		Filament Current "Amperes"	"A" Battery Supply "Volts"	Type *Amperite* to Use
Cunningham CX-301A C-301A	Radiotron UX-201A UV-201A	.25	6	*Amperite* No. 1A
Cunningham CX-300A CX-340	Radiotron UX-200A UX-240	.25	6	*Amperite* No. 1A
Cunningham CX-112 CX-371	Radiotron UX-112 UX-171	.5	6	*Amperite* No. 112
Cunningham CX-299 C-299	Radiotron UX-199 UV-199	.06	4 or 4½	*Amperite* No. 4V-199
Cunningham CX-299	Radiotron UX-199	.06	6	*Amperite* No. 6V-199
Cunningham CX-220	Radiotron UX-120	.125	4 or 4½	*Amperite* No. 120
Cunningham C-11 CX-12	Radiotron WD-11 WX-11	.25	1.5	*Amperite* No. D-11
West. Elec. 216A	Any 1 Amp. 5 V. Tubes	1.0	6	*Amperite* No. 1
Magnatron DC-201A	Any ¼ Amp. 5 V. Tubes	.25	6	*Amperite* No. 1A
Magnatron DC-112 Super-Rex	Any ½ Amp. 5 V. Tubes	.5	6	*Amperite* No. 112
Magnatron UC-120	Any .12 Amp. 3 V. Tubes	.125	4 or 4½	*Amperite* No. 120
Schickerling SX-8100	Any .06 Amp. 3 V. Tubes	.06	6	*Amperite* No. 6V-199
De Forest DV-2 DV-5	Schickerling S-4000 S-700	.25	6	*Amperite* No. 1A
De Forest DV-3	Any .06 Amp. 3 V. Tubes	.06	4 or 4½	*Amperite* No. 4V-199
Any ¾ Amp. 5 V. Tubes	Any ¾ Amp. 5 V. Tubes	.75	6	*Amperite* No. 3A
Ceco A-201	Stewart Warner SW-501A	.25	6	*Amperite* No. 1A

Fig. 184. Table Indicating What Size Amperites to Use on Various Tubes in General Use

casionally forget to turn off the filament circuit of the receiver, after a program is finished, with the result that the tubes are lighted over night or perhaps for several days. To warn the owner that the tubes are lighted, the use of a pilot lamp is convenient, two types of which are shown in details 1 and 2 of Fig. 185. The pilot lamp is provided on the panel, so that it can readily

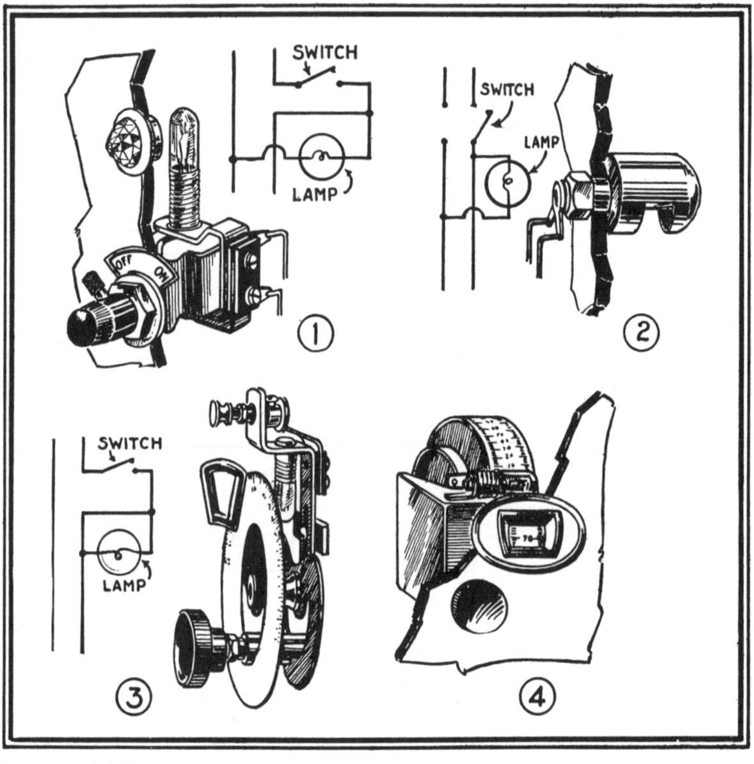

Fig. 185. (1) Filament Switch and Pilot Lamp Combined. (2) Separate Lamp Which Serves to Illuminate Panel Controls. (3) and (4) Illuminated Dials of the Latest Types. The Diagrams Show how the Lamps and Switches Are Wired

be seen. It is hooked up on the same side of the switch as the tubes, and is wired in parallel with them, as shown in the diagram. Dials having lamps incorporated to provide illumination for reading them are hooked up in a similar manner, as shown in details 3 and 4 of Fig. 185. It is not advisable to use pilot lamps and dial lamps on receivers drawing their filament current from dry cells, but they will be found to be of considerable utility on receivers operating on storage A-batteries or A-eliminators.

The Tuning Circuit and Its Troubles

Every receiver has a tuning circuit of some kind. The radio currents picked up on the aerial or loop are conducted to the tuning device, which by means of variations in inductance or capacity, brings the tuning circuit in resonance with the incoming wave.

Tuning Device Short-Circuited or Open-Circuited.—Fig 186 shows a coil of a tuned radio-frequency circuit. Such a coil must

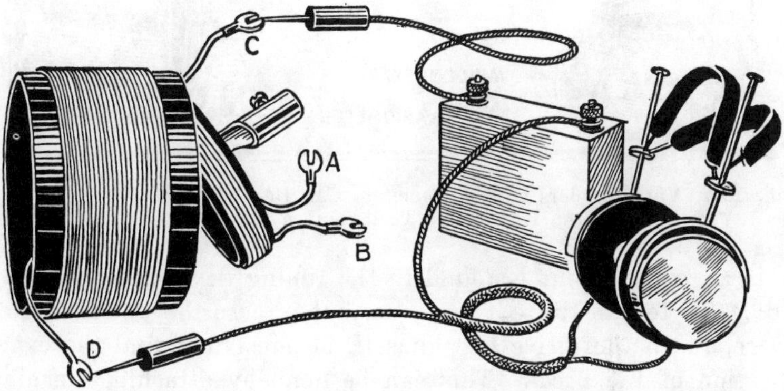

Fig. 186. Using the Headphone Tester in Series with a Small C-Battery to Make Continuity Tests on Tuning Coils and Air-Core, Radio-Frequency Transformers. The Absence of a Click Indicates an Open Circuit

be tested for continuity between points AB and between points CD. The tuning condenser is disconnected from the coil, and is tested for short circuits. Examine the wiring between the parts and see that all the wires are intact and make good electrical connection.

Variometers and vario-couplers are not used in receivers as much as they were a few years ago. However, if this method of tuning is employed, make a continuity test on the variometer and vario-coupler as indicated in Fig. 187. A variometer has a stator and a rotor (stationary and movable) coil, which are connected in series. The connections between the two coils may have been twisted off, or one winding may be rubbing against the other, fraying off the insulation, and shorting out parts of the windings.

A vario-coupler is similar to a variometer, but differs from it in that the windings are separated. Both windings should be tested separately for continuity, and the flexible connections should also be tested for open circuits.

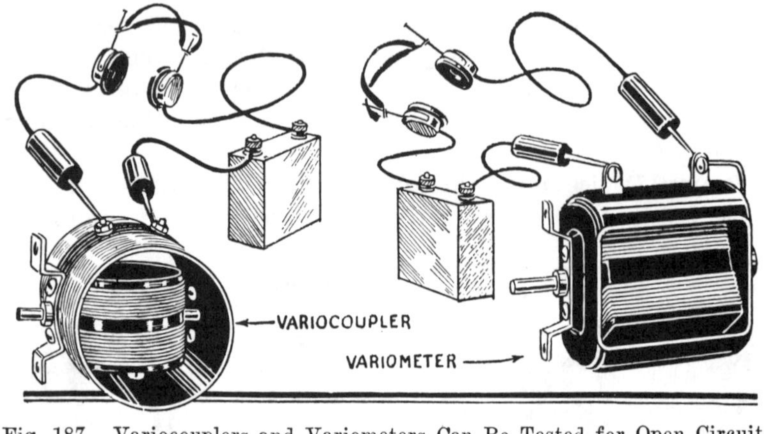

Fig. 187. Variocouplers and Variometers Can Be Tested for Open Circuits with a Headphone Tester and a C-Battery

If an open circuit is found in the tuning device, it is usually advisable to substitute a new one, but in case the owner wishes to repair the defective one himself, he must first find the exact location of the break. This can be done by attaching darning needles to the testing tips with tape, and then pushing the sharp points through the insulation of the coil. Start at the ends and progress toward the center until a click is heard. By this method the break can be accurately located. To repair it, scrape off the insulation from the wire on either side of the break and twist the ends together, after which solder is applied, so that the electrical connections between the broken ends will be restored.

The flexible connections on tuners often cause trouble as the stranded wire becomes brittle and breaks, resulting from constant twisting, and sometimes it is even pulled off. Never use solid wire when replacing this flexible lead, for the same trouble will then reoccur within a very short time. Always use a piece of well-insulated, stranded wire, preferably double-covered silk insulated wire, and make a good soldered connection at each end.

Wrong Connections.—If a receiver happens to be of the home-made variety, one should trace the tuning circuit carefully to see whether or not it corresponds to the wiring diagram. Wiring mistakes often occur, and wrong connections may also be found in the tuning circuit. If the receiver is of the regenerative type, employing a tickler coil for regeneration, one may find that the connections are simply reversed, so that the feed-back current runs through the tickler coil in just the opposite direction to the flow of the intercepted radio current through the tuning coil as shown in Fig. 188. The purpose of a tickler coil is to superimpose

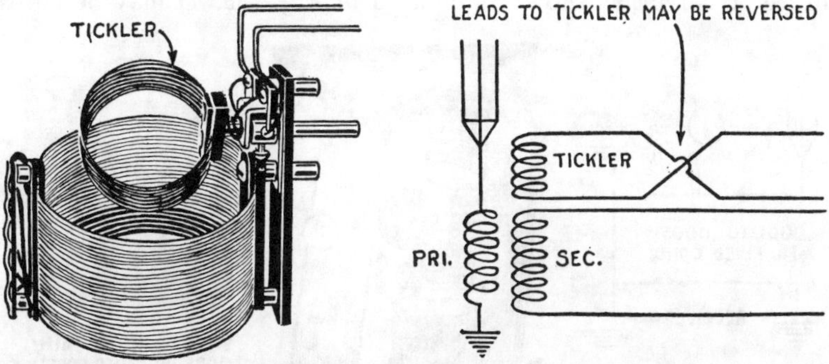

Fig. 188. Left: Low-Loss, Three-Circuit Tuner on Which the Tickler Coil Can Be Rotated. Right: Diagram of a Three-Circuit Tuner in Which the Leads to the Tickler Have Been Reversed

a comparatively strong current on the weak incoming radio current by means of induction, and, accordingly, the feed-back current must be fed through the tickler coil in the same direction as the incoming signal, instead of in opposition to it. Some tickler coils are able to turn the full radius of 360°, and by turning it around half-way one can readily tell whether regeneration becomes greater or less, and such an adjustment may serve the same purpose as reversing the leads to the tickler. However, do not strain the flexible leads to obtain this adjustment; it is better to reverse them.

Broad Tuning.—When a receiver tunes too broad, the trouble usually is too much inductance in the tuning circuit, which includes the aerial. The inductance of a conductor is proportional to its length, and therefore the more wire a tuning coil has, the greater will be its inductance value. Broad tuning usually means considerable interference owing to station signals overlapping, and it is sometimes so bad that several stations can be heard simultaneously. This is especially true in cities where a number of stations on different wavelengths may be broadcasting at the same time. In farming communities, remote from broadcasting stations, less trouble is experienced from this source, except when two stations on the same or nearly the same wavelength are tuned in.

The first step in eliminating the trouble is to decrease the length of the aerial. If shortening the aerial does not help sufficiently, connect a small fixed condenser of from .0001 to .0005-mfd. capacity in series with the aerial and the aerial binding post on the receiver, as shown in detail 1 of Fig. 189. It may be neces-

Fig. 189. Three Methods of Increasing Selectivity: (1) Insertion of a Small Fixed Condenser in the Aerial Circuit, (2) Shielding an R. F. Stage, (3) Variable Primary on Coil to Obtain Loose and Closed Coupling

sary to experiment a little in order to find just what value of condenser is most effective for this purpose. Such a condenser may be connected in either the aerial or the ground circuits; many prefer the latter method, claiming that it is more effective. If this does not help sufficiently, try an inside aerial, which is usually much more selective than an outdoor aerial.

Where extreme selectivity must be had, additional methods usually are necessary. The tuning circuit should be completely shielded in a brass, copper, or aluminum box or can, and the can must be grounded to the A-negative line of the receiver. The purpose of such shielding is to eliminate all pick-up of signals resulting from the tuning coil itself acting as an aerial. It may be noticed in many tuned radio-frequency receivers, located in the vicinity of a broadcasting station, that reception can often be had with the aerial and ground leads entirely disconnected; this is an instance where signals are picked up directly by the coils in the receiver, and broad tuning is usually an accompanying evil. Shielded coils do not pick up atmospheric "strays," and the trend in receiver design is toward the use of "canned coils" entirely. Detail 2 of Fig. 189 shows an aluminum box which is used for shielding the tuning circuit. It is large enough to hold a variable condenser, a coil, and a tube. Shielding radio-frequency stages also eliminates inter-stage coupling.

Sometimes it is necessary to alter the coils to some extent in order to gain selectivity. If the coil is closely coupled, that is, if the primary and secondary windings are in the same plane, and as close together as possible, the tendency of the coil will be to tune broad. However, the closer the coils are placed together the greater will be the transfer of energy. By mounting the primary winding on a pivot of insulated material, to permit it to be turned freely in and out of the plane of the secondary coil, as shown in detail 3 of Fig. 189, selectivity can be gained. This method has the advantage of more transfer of energy, when this is needed, as, for instance, on "silent night," when the receiver is used to tune in distant stations, and less transfer of energy, with greater selectivity, when several nearby stations are broadcasting.

If the primary winding is permanently fixed, and one does not wish to go to the trouble of making it variable, two other methods can be used to gain selectivity. One of these is to remove the primary winding and substitute a smaller one, having the same number of turns, but wound to such a diameter that there will be more space between it and the secondary winding. The other method consists in reducing the number of turns on the primary winding.

Broad tuning can also be caused by the antenna effect of battery eliminators. This trouble and its cure is treated in detail under the section entitled, "Curing Antenna Effect of Battery Eliminators."

Wave Trap to Reduce Station Interference.—When the receiver is located in the vicinity of broadcasting stations, interference is often experienced from two stations operating on wavelengths only a few meters apart. A receiver located two or three blocks from a station picks up the signal from this station over a wide range on the tuning dials, and it can be heard in the background even when other more distant stations are tuned in. In such cases the receiver should be entirely shielded, a short aerial should be used, and if the trouble still persists, a wave trap should be provided. A wave trap is a device placed ahead of the receiver, and is usually connected in the aerial line. The trap consists of variable inductance and capacity devices, which can be tuned to any point on the waveband. The function of a wave trap is to absorb the signal of an undesired station, so that stations operating on different wavelengths can be brought in without interference. It is possible, however, to absorb only one signal or several signals on the same wavelength, at one time, and if there are two interfering stations on different wavelengths, which prevent the satisfactory reception of a third station, only one of the interfering stations can be tuned out with a wave trap. For instance, if the wave trap is adjusted to absorb a signal having a 425-meter wavelength, all stations at this wavelength will be unheard, but interfering stations at 410 meters may still cause interference. Some wave traps are more efficient than others, as they enable the operator to tune them more closely. There are two common faults with a great number of wave traps; they do not eliminate the interfering signal effectively or they decrease the volume of other signals to an appreciable extent.

How to Make an Efficient Wave Trap.—One type of wave trap, which has been found to be highly efficient, is shown in Fig. 190. This type may be slightly more difficult to make than the average type of wave trap, but results will reward the time and work of building it, and its construction is within the reach of every radio fan who is able to use a few tools. The instruments needed to construct this wave trap are a .0005-mfd. variable condenser, a 75-turn tapped spider-web coil, a vario-coupler with a tapped stator winding, two nine-point inductance switches, and two binding posts. A neat little cabinet having a 6 by 7-in. panel can be picked up at almost any radio shop, as well as two large dials, similar in size and design, for the condenser and the vario-coupler. The smaller dials and knobs are furnished with the inductance switches. When selecting a condenser, get one that can be readily mounted in the upper corner of the cabinet. Before marking off and drilling the panel, carefully lay out the instruments on the back side, so that they will not be crowded or come too close to the sides and top of the cabinet after the panel is screwed in place. The spider-web coil is screwed to the baseboard and its taps are soldered to the lugs on the first inductance switch, in consecutive order, as shown in the diagram. The taps on the stator, or outer winding of the vario-coupler, are similarly soldered to the lugs of the second inductance switch. The levers of both switches are connected to the binding posts nearest them,

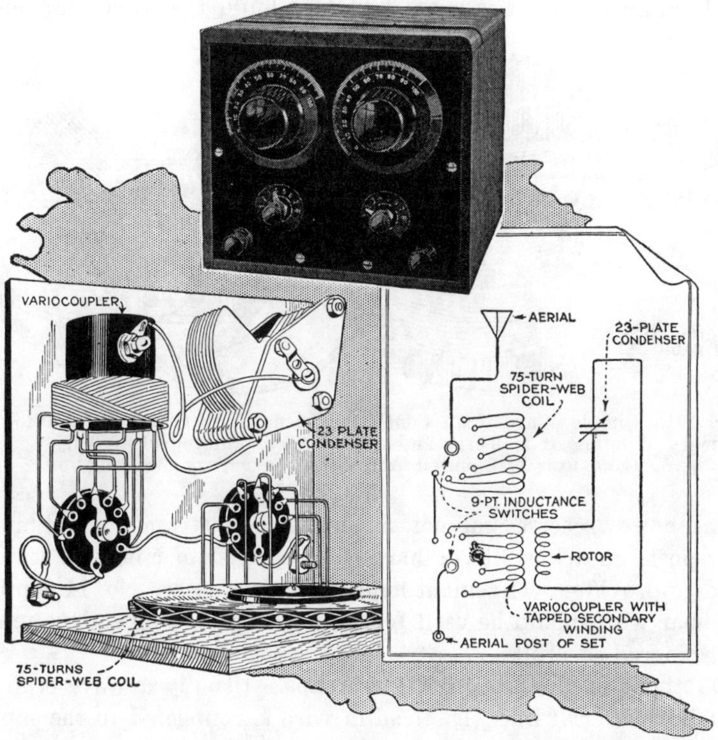

Fig. 190. Homemade Wave Trap Which Consists of a Variocoupler, Variable Condenser, Tapped Spider-Web Inductance Coil, and Two Inductance Switches. It Enables One Completely to Tune out an Interfering Station

(Courtesy of "Popular Mechanics Magazine")

and the last tap on the first switch is connected to the first tap on the second switch. The rotor of the vario-coupler is wired to the condenser, and the wave trap is then ready for use.

If the radio owner wishes to make a less elaborate wave trap, one can readily be made by winding two coils of No. 26 enameled or double silk-covered copper wire, on a cardboard tube about 3½ in. in diameter, as shown in Fig. 191. About 45 turns of wire are first wound on the tube, the ends being fastened securely so that they will not unravel. Then apply a winding of tape in the center over which a small coil of 15 turns of wire is wound. This coil should also be securely held in place with a layer of tape wrapped over it. A .0005-mfd. variable condenser is shunted across the 45-turn coil as indicated, and the ends of the small coil are connected to binding posts. It is a good idea to build the wave trap in a

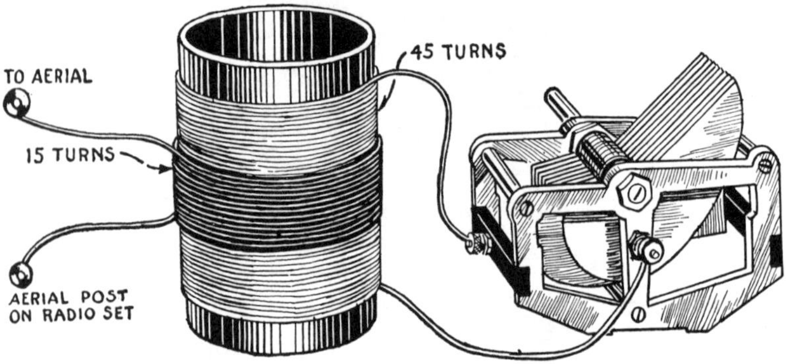

Fig. 191. Simple Wave Trap Consisting of an Inductance Coil Having a Primary Winding of 15 Turns and a Secondary Winding of 45 Turns. The Condenser Is Shunted Across the Secondary Winding

small box or receiver cabinet, mounting the instruments and binding posts on a bakelite or hard-rubber panel to minimize losses, and to provide a convenient holder for the device. No. 14 copper bus-bar wire should be used for wiring, and all connections must be soldered carefully.

Methods of Connecting Wave Traps.—Usually a wave trap is cut in the aerial line. The lead-in wire is connected to the input terminal of the wave trap, and a separate wire connects the output terminal of the wave trap to the aerial binding post on the receiver. In some cases it is, however, necessary to connect the wave trap in a different manner. On radio-frequency receivers, on neutrodyne receivers, and on receivers having reflex circuits, a wave trap should be connected to the grid of the last radio-frequency transformer and the grid of the detector tube, just before the grid leak, as shown in Fig. 192. If a receiver has a regenerative circuit, such as the ultra-audion, the trap can be cut in either the aerial or the ground circuit, or it may be found that better results are obtained if it is connected in shunt across the aerial and ground, as indicated in Fig. 193, detail 1. In some cases it may also be found effective to connect an extra .0005-mfd. variable condenser in series with the aerial and the wave trap, if the latter is connected in the aerial line, or between the ground and the wave trap, if the latter is cut in the ground line. Detail

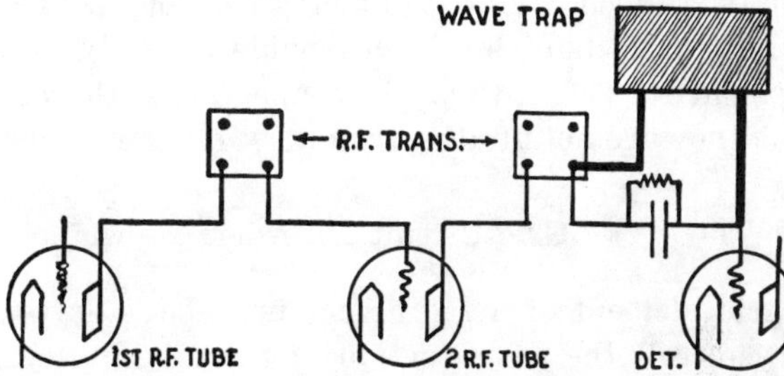

Fig. 192. Method of Connecting a Wave Trap Between the Grid Terminal of the Last R. F. Transformer and the Grid of the Detector Tube

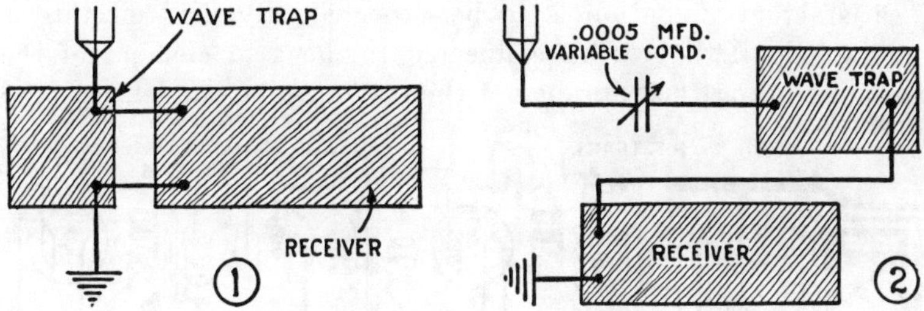

Fig. 193. (1) Shunting a Wave Trap across the Aerial and Ground Terminals of a Receiver. (2) Method of Connecting a Wave Trap in the Aerial Line of Receiver with an Extra Variable Condenser

2 of Fig. 193 shows how this is done. It may be necessary to experiment, to find out just what method is most effective on any particular receiver.

How to Operate a Wave Trap.—After a wave trap is hooked up to the receiver, tune in the signal of a nearby broadcasting station. Carefully adjust the dials on both the receiver and on the wave trap so that this signal will be brought in at its maximum volume. Then turn down the rheostats so that the signal can just be heard. Tune again on the dials to bring up the volume of the signal as much as possible without adjusting the rheostats. After this has been done, the rheostat is advanced a little for comfortable volume, and then the control of the wave trap is manipulated until the signal is completely absorbed, or nearly so. When the signal is inaudible, the receiver can be tuned in to any other station except those which have the same wavelength as the trapped signal. It will be found that the use of a wave trap may alter the dial readings to some extent. It should be remembered, when using a wave trap, that, although an undesired signal can be completely trapped, its harmonic may be brought in at another point on the tuning dials, but such trouble is rarely experienced. Wave traps should be used only in connection with vacuum-tube receivers as they are not at all satisfactory for use on crystal sets.

Grid-Circuit Disorders

As the grid element of a radio vacuum tube is extremely sensitive and controls the entire action of the tube by letting more or less plate current pass in synchronism with the fluctuations of voltage applied to it, the grid circuits must be in good condition if satisfactory reception is to be expected. By grid circuits is understood all the wiring connected to the grid element of the tubes, including the windings of the instruments, condensers, and

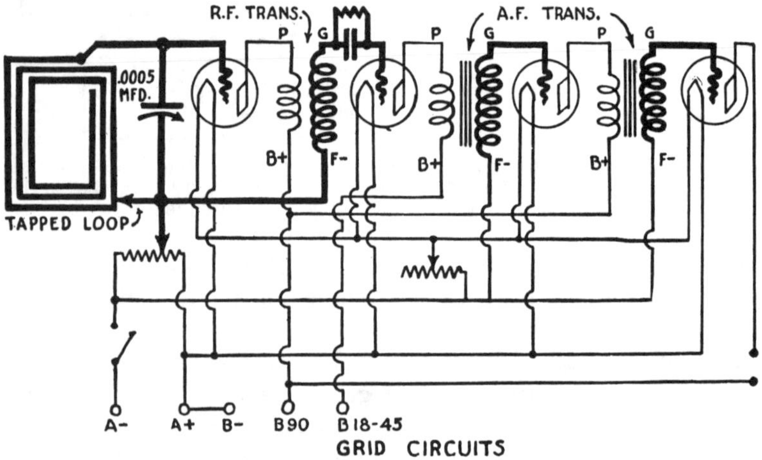

Fig. 194. Diagram of a Typical 4-Tube Loop Receiver, the Grid Circuits Being Indicated by the Heavy Lines

coils cut in this wiring. Fig. 194 shows all the grid circuits of a typical receiver. The most common troubles encountered in grid circuits are high-resistance joints and open circuits, faulty connections, short circuits, grounds, induction troubles and correct values of grid leak and condenser, and faulty by-pass condensers.

High-Resistance Joints and Open Circuits.—High-resistance joints and open circuits may be found in the wiring, owing to a broken joint, which has been poorly soldered, a flux-insulated or a corroded joint. Such defects can sometimes be found by testing with a headset and a C-battery. "Cold"-soldered joints, which can be detected by their rough and ragged appearance, should be resoldered, using paste flux. The soldering iron should be kept on the joint until all the solder is melted and has a smooth surface as soon as the iron is removed. Flux-insulated joints can readily be wiped clean with a rag saturated in alcohol. The practice of wiping soldered connections in this way should be observed when any work of this kind is done. High resistance, caused by corrosion, may result if all the flux is not wiped off, for most flux contains a certain amount of acid, which eats down into the wire and leaves an insulating crust. Paste flux may be used in radio receivers, although rosin flux is preferable, as it has no corrosive effect. Acid flux should never be used in radio receivers under any circumstances, for it has an exceptionally ruinous effect. Sometimes high-resistance joints resulting from the use of excessive flux and poor soldering can best be remedied by resoldering all the joints of the receiver, instead of trying to find the faulty ones.

Other Open Circuits in Grid Leads.—If the wiring is broken at any point, this can usually be seen at a glance, unless stranded, insulated wire is used. Open circuits must then be found with the headset tester. Broken or badly frayed wires should be replaced. Open circuits may also exist in the secondary winding of transformers, but this trouble is rare. In impedance circuits it is sometimes possible for the plate current to get into the grid circuits, if the impedance device is defective or if the grid condenser is short-circuited. Such conditions prevent the detector from operating, and one or more tubes may be burned out. A broken-down C-battery in a grid circuit is another cause of an open circuit.

Short Circuits and Grounds.—Bent plates of variable condensers may short out a grid circuit. This trouble is especially likely to occur in tuned radio-frequency circuits, where the condensers are shunted across the secondary windings of the coils. Rotors of inductance coils may touch the stators at some point when rotated, which results in scraping off the insulation, and then a short circuit will result as soon as the wires of the separate coils make electrical contact. Tubes and transformers can be burned out by such defects. If the receiver is transported or is handled frequently, the utmost care should be taken to prevent bending of the wires. Many receivers are wired with bare copper wire and some leads run quite close together, so that it takes only a little pressure to make a contact between them, or bend them down permanently. It is best practice to keep receivers in cabinets, or in suitable boxes or cartons, while moving them about.

When receivers are located in damp places, the insulation on the coils absorbs dampness from the air. Cotton has this tendency more than any other kind of insulation. Silk does not absorb a great amount of dampness, and enamel practically none at all. If the insulation of coils is saturated with moisture and the turns of wire are wound closely together, there will be a considerable loss of grid current, which results in loss of volume. In the case of a tuning coil connected directly to the ground and the aerial, the incoming signal will be partly grounded, for some of the current is short-circuited over the insulation, and passes on to the ground without going through the coil.

Inductance Troubles.—Uncontrollable whistling and howling in receivers is often caused by pick-up on grid leads. All electrical conductors through which pulsating currents flow, set up an electro-magnetic field around them, the lines of force of which are at right angles to the conductor. This field induces currents in other wires running through it. Plate leads in receivers often set up undesired currents in the grid leads in this way. To eliminate or minimize this trouble, the grid and plate leads should always be separated as far as possible, should not be run parallel, and should be kept as short as is consistent with good arrangement of the instruments. A plate lead near a grid condenser, even though the two are not parallel, can cause considerable trouble. Unshielded windings of coils and r. f. transformers can also pick up interference from a loop antenna placed near the receiver, from the lighting circuit or other sources.

Faulty Grid Leak and Condenser.—Defective grid leaks and condensers are likely to cause trouble. ''Motorboating,'' which is explained in a section under that heading, is one annoyance from this source. Squeals, howls and signal distortion can be caused by defective grid leaks and condensers. When the trouble is not found elsewhere, remove the condenser and test it thoroughly according to the instructions given on condenser testing. Holding a hot soldering iron on the terminals of a grid condenser sometimes ruins it, and for this reason it is best to solder the leads to soldering lugs separately, and then atach the condenser to the lugs by means of screws, as shown in Fig. 195.

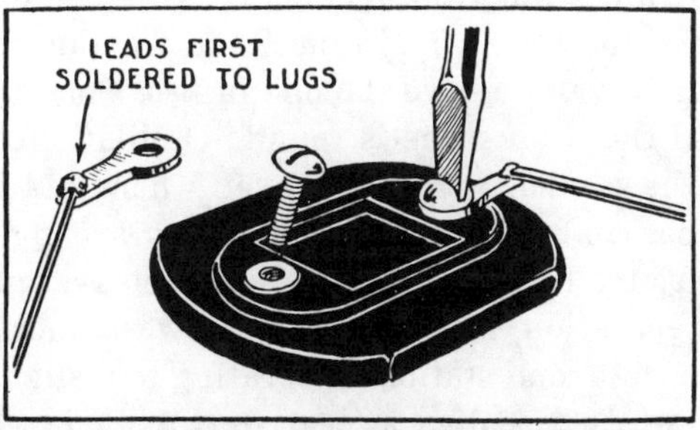

Fig. 195. Never Solder Wires to Lugs already Attached to Fixed Condensers. Screw the Lug to the Condenser after Soldering

Grid-Leak Value Too Low.—Uncontrollable oscillations and "hang-over" effects result from the use of a grid leak which has too low a resistance value. One may have noticed this effect when tuning in a signal. As the tuning dial is rotated to the point

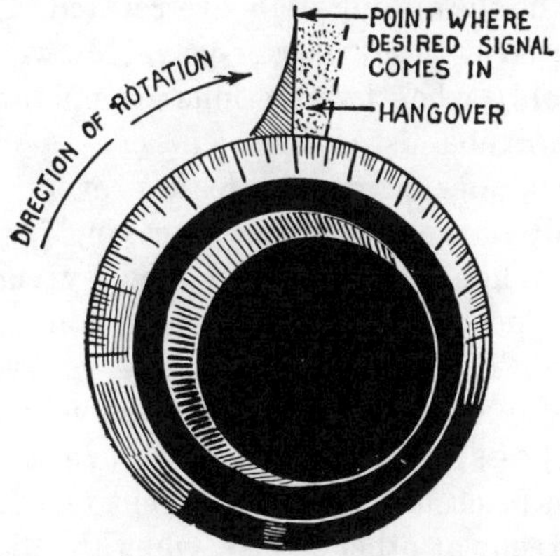

Fig. 196. Too Low a Grid-Leak Value often Causes a "Hangover" of Signals

where the desired signal comes in, the volume will increase as the exact point of proper tuning is reached, indicated by the cross-hatched section in Fig. 196. After the signal comes in at the proper tuning point on the dial, and reception is satisfactory, a steady whistle or howl may suddenly commence. It will keep on even if the dial is turned a few degrees from the station signal in the direction indicated by the arrow. Such trouble is called an oscillation hang-over, and is caused by the use of a grid leak having too low a resistance. The remedy is to substitute progressively higher values until the trouble ceases. The danger then is to use a leak of too high a resistance, and for this reason several leaks, ranging in value above the one in use, should be tried successively until the proper one is found. Oscillations of this kind, which have the tendency to hang over when rotating the dials should not be confused with oscillations arising from other sources, such as too high grid-leak values, incorrect plate voltages, too much regeneration, inter-coupling between successive stages, and interference from stations operating on slightly different wavelengths, which is called a "carrier-wave" howl.

Grid-Leak Value Too High.—If the grid leak has too high a resistance it may cause extremely harsh-toned reception, a constant series of clicks usually termed "motorboating," howls and squeals or complete inoperation of the receiver. A howl, due to this cause, is not the steady howl like a hang-over oscillation caused by too low a grid-leak value, but the pitch of this howl changes slightly as the tuning dials are rotated. It may, however, readily be confused with a "carrier-wave" howl. A carrier-wave howl is evident only when two stations are operating at the same time and their wavebands conflict.

Too much resistance in a grid leak prevents the dissipation of excessive grid charges, which is necessary for the action of a detector tube. When the grid is choked by such an excessive charge, there are no grid fluctuations, and consequently the tube will not operate. The trouble can easily be overcome by substituting grid leaks of lower values successively, until the squeal ceases. As progressively lower values are tried, the squeal will often rise in pitch until it becomes entirely inaudible. If the squeal does not recur at other points, when the dial is rotated, the value used is correct, but if the squeal is evident at other points on the dial, still lower values should be used.

Variable Grid Leaks.—After the proper value of grid leak has been found, this value should be kept, unless of course, another type of detector tube is used, and in that case grid-leak readjustment may be found necessary. Keeping a grid-leak value constant, when once adjusted, requires the use of good stable resistors, and variable grid leaks usually cannot maintain exact adjustment. Some variable grid leaks make use of carbon granules for resistance variation. They can be compressed to various degrees of compactness by means of a screw, in order to provide any desired resistance. However, owing to the gradual settlement of the granules, the degree of compression, and consequently the resistance value, will change. The tighter the screw is turned, the lower the resistance will be, and vice versa. Other types of variable resistances may be mechanically better suited to the purpose but, where a variable control is provided, the radio owner is tempted to tamper with it continually, in order to improve reception. Fixed resistors of the correct values are preferable, as they are usually left untouched.

"Click" Tests on Grid Circuits.—Tests with a headset and C-battery indicate many open circuits, but the delicate grid circuits may have faults which cannot be detected by the headset test. One method of locating troubles in grid circuits is to touch the grid and negative-filament terminals of the tube sockets with the index finger and thumb of one hand as shown in Fig. 197.

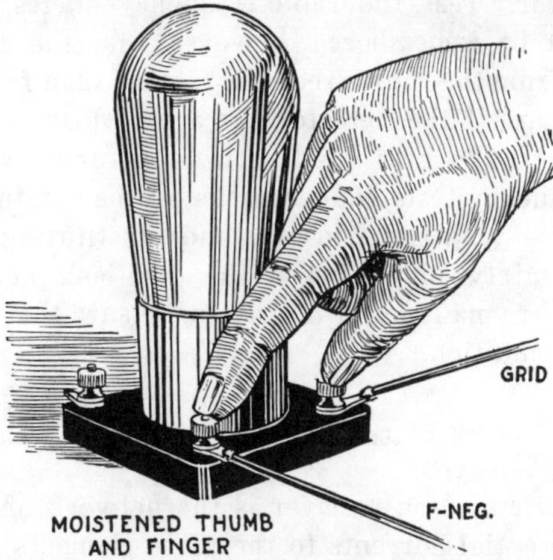

Fig. 197. Testing the Grid Circuits by Touching the Moistened Thumb and Index Finger to the Grid and F-Negative Terminals of Sockets

Both the thumb and index finger should be moistened when making this test. The test must be made while the receiver is hooked up and the tubes are lighted, so that a click will be heard in the phones or loudspeaker each time the fingers are touched to the above-mentioned terminals. Start the test on the last audio tube; if no click can be heard, there is an open circuit in the output lines to the loudspeaker, an open-circuited or disconnected output transformer, a broken phone cord, or a defective loudspeaker or phone. Trouble indicated by such grid tests is not necessarily confined to the grid circuits, but may exist anywhere between the point tested and the loudspeaker. If the last audio stage seems to be in good condition, as indicated by a click, test the preceding audio stage in the same way. The absence of a click here indicates trouble between the last two stages of audio; it may be an open circuit in the wiring or a defective transformer, resistance, or impedance device, depending on the particular kind of audio coupling used in the receiver. Resistance and impedance-coupled circuits have three stages of audio amplification, which can all be tested in the same way. When the grid condenser is touched, a squeal or howl can usually be heard, depending on the adjustment of the rheostat. It is a good idea to make this test first, so that one can determine whether the trouble is located in the audio end of the receiver or in the radio-frequency end, which is ahead of the detector. Test the radio-frequency stages in the same way. It must be remembered, however, that the clicks will be much fainter from the radio-frequency stages than from the audio-frequency stages, if the detector has a grid leak and condenser. By shorting out the grid condenser temporarily, while making the radio-frequency tests, louder clicks can be obtained. This can be done by removing the grid leak and substituting a paper clip or a closed safety pin between the grid-leak holders. When "click" tests are made on grid circuits, be sure that all the tubes are in good condition.

Plate-Circuit Troubles

The plate circuit of a receiver is the network of wires which carry high-potential currents to the plate elements of the tubes. Fig. 198 shows the plate circuit of a typical receiver. It will be noticed that the negative plate line is connected to the negative filament line, in order to reach the filaments of the tubes. As the grid circuits are all connected to the A-negative line, they also carry the B-negative supply, although they do not function in the plate circuit and are not included in it, as indicated in the diagram. All instruments and parts of windings of instruments through which the plate circuit flows belong to it, and troubles are often located there, especially in the case of transformers. A large percentage of receiver troubles are found in the plate circuits, and this is quite natural as this circuit must stand up under high voltages. Plate circuits are furthermore the cause of trouble in

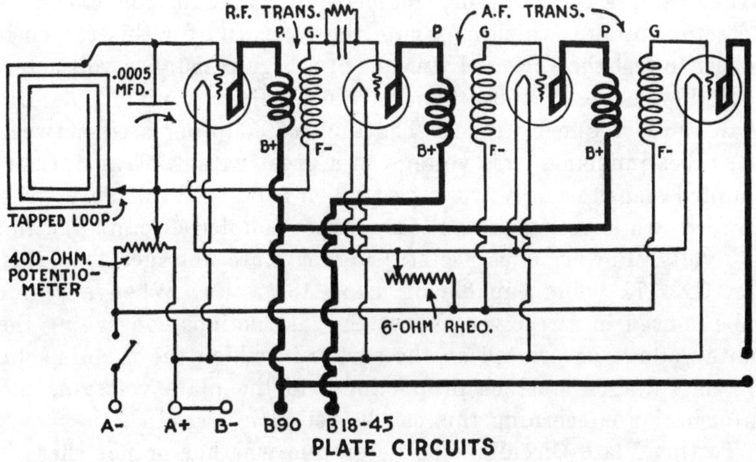

Fig. 198. Diagram of a Typical 4-Tube Loop Receiver, the Plate Circuits Being Indicated by the Heavy Lines

other circuits, as for instance when there is a short circuit, which permits the plate voltage to get into other circuits. Induction from the plate circuit, evidenced by squeals and howls, is also common.

Incorrect Plate Voltages.—Improper operation, harsh-toned reception, and paralyzed tubes can often be traced to wrong plate voltages. Many radio fans disregard the instructions accompanying the receivers and find that they can obtain more volume and greater distance by increasing the plate voltages, especially on the audio tubes, but the payment for such reception is decreased tube life, distorted signals, harsh-toned reception, and open-circuited transformer and loudspeaker windings. If the plate voltages on the radio-frequency tubes is excessive, uncontrollable oscillations may result, especially in radio-frequency receivers. It is advisable to keep these voltages as low as consistent with the proper operation of the receiver. Values of 67½ volts and 45 volts for radio-frequency tubes will often eliminate squeals and howls when tuning, provided the filament rheostats supplying these tubes are not turned up too high. Detector-plate voltage should not exceed 45 volts on tubes generally used for detection, and lower values than this will usually give still better results. "Microphonic" detectors can often be cured by simply reducing their plate voltage. Harsh-toned reception is often due to excessive detector voltage. When using a B-eliminator, trouble is apt to result from excessive detector voltage, which may ruin the tube, and for this reason it is best to test the voltage by means of a high-resistance voltmeter.

Audio-frequency tubes generally use a higher plate potential than radio-frequency tubes. The kind of coupling used between the tubes influences this voltage to a great extent. Transformer-coupled stages usually work best when a potential of 90 volts is applied, while resistance and impedance-coupled circuits demand 135 volts. Power tubes use still higher plate voltages than this, the UX 171 being capable of using 180 volts. When a power tube is used in a receiver, follow the instructions concerning the plate voltage as printed on the carton in which the tube is sold. C-bias voltages must be proportional to the plate voltages, and information concerning this can be found in Fig. 153.

Testing Plate Circuits.—To determine whether or not there is trouble in the plate circuit, several simple tests can be made. First of all, turn on the filament switch to light the tubes, the plate supply being connected. While the receiver is thus in operation, plug the loudspeaker into the output jack. When this is done a distinct click should be heard, which indicates that the plate circuit of the last audio tube is in good condition. If the receiver has no output jack, but the loudspeaker is connected to binding posts on the subpanel, as is often the case with receivers installed in consoles, then one of the loudspeaker terminals is disconnected and tapped on the binding post once or twice. This should give the same kind of click. If the receiver has a separate jack for the first audio stage, the loudspeaker plug is inserted into it in the same way to cause a click, but this click is normally weaker than the one obtained when plugging into the last jack.

After the last audio stage has been tested and found to be in good condition, the wire that supplies the audio plate voltage is disconnected. This can be done either at the receiver or at the batteries. The end of the wire is then touched to the terminal to which it was connected. A click and a spark should be observed each time this is done. The first spark is usually the largest, because of the charge of electricity that has collected on the wiring and instruments in the receiver. If a spark is observed but no click is heard, then there is a short circuit, which shunts out the loudspeaker and one or more instruments. If no spark can be seen, there is an open circuit in the wiring, which can be located by headset tests. This "spark-and-click" test tells whether or not the plate circuit in the audio stages is functioning. If no defect is found, connect the battery lead to its proper terminal, disconnect the detector-plate lead, and make the same test, to determine whether there is any trouble in the detector-plate circuit. Test the radio-frequency plate circuits in the same way.

Variable Control of Plate Voltage.—With the use of ballasts in receivers instead of rheostats, the variable ratio between the filament voltage and plate voltage is eliminated. In radio-frequency

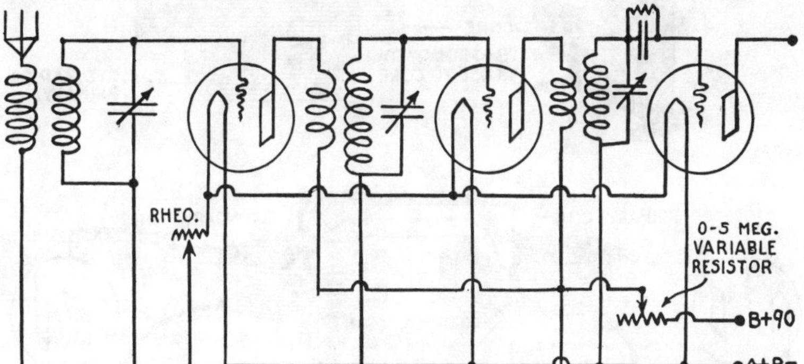

Fig. 199. Method of Controlling the Plate Voltage of Radio-Frequency Tubes by Means of a Variable Resistor Connected in the B-Positive Line Supplying 90 Volts, as Indicated

receivers using ballasts, oscillation is often caused by an incorrect ratio between these voltages. One method of overcoming this trouble is to provide a variable resistor in the plate circuit of each tube as indicated in Fig. 199, which shows such a method of control for radio-frequency stages. In this way tubes can be operated without oscillation squeals, and there will be considerable improvement in reception. Such a resistor connected in the detector-plate circuit will help to tune in distant stations more easily.

Faulty Instruments in Receiver

Radio troubles in the receiver usually manifest themselves in the individual instruments, and the owner should therefore acquaint himself with the action of the instruments, how to operate them properly, how to maintain them in order to prevent trouble, and, in case of trouble, how to remedy it.

A. F. Transformer Construction and Operation.—The oldest form of coupling between tubes is transformer coupling. A transformer consists of two coils, a primary and secondary, wound on a core, as shown in Fig. 200. The current in the primary winding

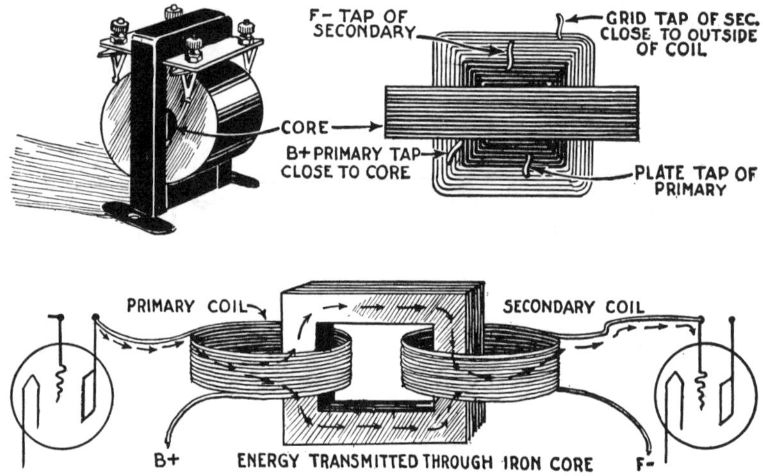

Fig. 200. How Audio-Frequency Transformers Are Assembled and How They Work. The Current Flowing through the Primary Coil Induces a Magnetic Flux in the Iron Core, Which in Turn Induces a Current of Higher Voltage in the Secondary Coil

induces a stronger current in the secondary winding by electromagnetic induction, which is concentrated in the iron core. This step-up of voltage depends entirely on the number of turns in the windings. If the secondary has twice as many turns as the primary, the voltage will be doubled; if it has three times as many turns the voltage will be tripled, etc. The ratio between the primary and secondary windings of audio transformers varies, the 1 to 2, 1 to 3 or 3½, and the 1 to 5 or 6 ratio being the most common. Unless properly designed, transformers may cause considerable distortion of music and speech, and a transformer of poor design may be totally unable to reproduce low-frequency signals. This is due to the fact that a transformer usually works better at one frequency than another, although some manufacturers claim that their transformers have various points of resonance to the frequency on which these transformers operate best, and this makes for more even amplification. Audio frequencies range from 20 to 10,000 vibrations per second, and the best transformer is one which amplifies more or less uniformly over the entire range. More uniform amplification can be obtained from most transformers by connecting a variable resistance unit having a range of from 0 to 5 megohms across the secondary terminals, as shown in Fig. 201. This practice will reduce the volume, but will improve the tone quality of the receiver. If the receiver has more than one audio transformer, provide such a variable resistor across the secondary of each transformer in order to get best results.

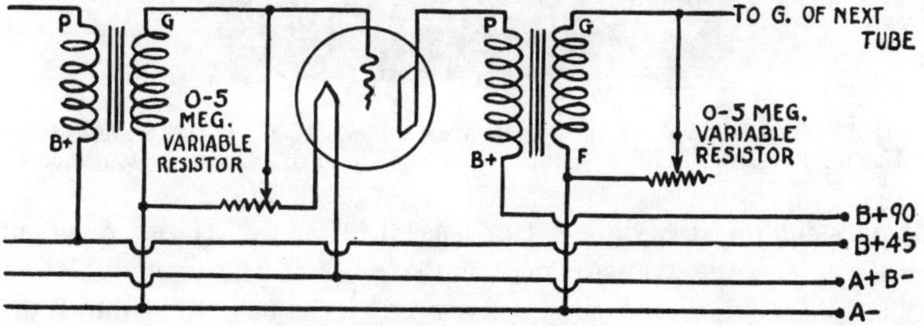

Fig. 201. Using 0-5-Megohm Variable Resistors across the Secondary Windings of Audio-Frequency Transformers in Order to Obtain More Uniform Amplification

Transformer Distortion.—In an ideal transformer the fluctuations between the input and output currents should be practically in perfect synchronism. However, in most transformers such action is impossible, owing to the reluctance of the core in transferring the energy. The result is a peculiar "lag" effect, particularly noticeable in cheap transformers which have poor core material and are not properly designed. The lag is characterized by one note overlapping into the next. Incorrect core value and material also give rise to distortion and harsh-toned reception.

Distortion in transformers can sometimes be traced to leakage through saturated insulation, coupling effects from other transformers, grounded secondary windings, and incorrect ratios of the windings. Moisture in the windings permits a loss of energy, which may develop a short circuit. Saturated coils often cause hissing and frying sounds in the loudspeaker or phones. Coupling effects between transformers are often in evidence where the transformers are of the open type, having no case or shell, and are located close to each other, as shown in Fig. 202. Open trans-

Fig. 202. Small Receiver in Which Unshielded A. F. Transformers Were Used. The Dotted White Lines Show the Field of Mutual Inductance

formers should always be set at least 3 in. apart. If this does not help, replace the transformers with those of the enclosed type. When shielded transformers are used, it is best to ground the case or shell to the negative filament line of the receiver. This can readily be done by removing some of the paint or lacquer from the shell at the point where it is held down on the subpanel with a screw. A soldering lug is slipped on the screw, from which a connection is made to the negative line, as shown in Fig. 203. In receivers where high-ratio transformers are used, distortion of signals is often noticeable. Under ordinary operating conditions it will be found that transformers having a 3 or 3½ to 1 ratio will be entirely satisfactory. If a better tone quality is desired, use 2 to 1 ratio transformers. On receivers where greater volume is needed, it is sometimes advisable to use a 5 or 6 to 1 ratio transformer in the second audio stage. Deep, hollow sounds are often caused by excessive capacity in a transformer, and this trouble can be remedied only by replacing the offending transformer. There are on the market a few devices which are essen-

Fig. 203. Shell of an Audio-Frequency Transformer Grounded to the A-Negative Line

Fig. 204. A Three-Stage, Audio-Frequency Amplifier Unit, Connected to the Output of the Detector, Improves the Tone Quality of an Old Receiver

(Courtesy of "Popular Mechanics Magazine")

tially amplifier units and are designed to be connected to the detector-plate output of any receiver. A unit of this kind is shown in Fig. 204.

Push-Pull Amplifier Reduces Distortion.—Audio-frequency amplification beyond two stages is not desirable, as it gives rise to considerable distortion of signals. However, greater volume without distortion can be obtained by means of a stage of "push-pull" amplification, this being substituted for the second audio stage. Fig. 205 shows a diagram of a push-pull amplifier. It consists of two special transformers, two tubes, two sockets, and a rheostat. Both halves of the alternating-current waves passing through the transformers are utilized, and the two tubes share the load equally, one balancing the other. The increase of amplification of a push-pull amplifier is not equal to two additional stages of audio amplification connected in the usual way; the increase in volume is only a little greater than that obtained

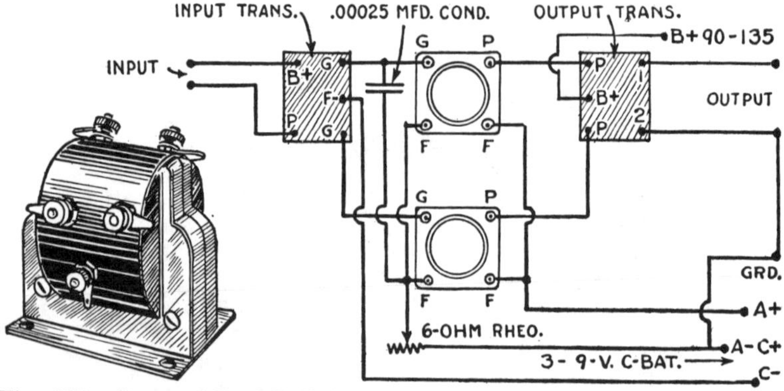

Fig. 205. Stage of Push-Pull Audio-Frequency Amplification Which Gives Maximum Volume with a Minimum Distortion of Sound. Two Tubes and Two Specially Made Transformers Are Used in This Hookup

from one stage of audio, but the undistorted reception is its chief advantage.

Open Circuits and Short Circuits in Transformers.—Transformers can be tested for open circuits, short circuits, and grounds by using the headset tester, as shown in Fig. 206, connecting a 22½-volt B-battery in series with the phones. The testing tips are first applied across the primary terminals, marked P and B+. A sharp click should be heard, which shows that there is continuity in the winding. The same test is applied across the secondary terminals, marked G and F—. It must be remembered that the click across the secondary terminals will be weaker than that across the primary terminals, because of the greater resistance of the secondary winding. If no click is heard, the winding being tested is broken or open-circuited.

A test should also be made to determine whether or not there is any leakage from one winding to another. When making this test, as indicated in Fig. 207, be careful to prevent the testing tips from touching each other. The tips should be held immovably to one of the primary binding posts and one of the secondary binding posts. Be sure not to rub the tips on the posts, for this may result in an incorrect test. In case a click is heard, both

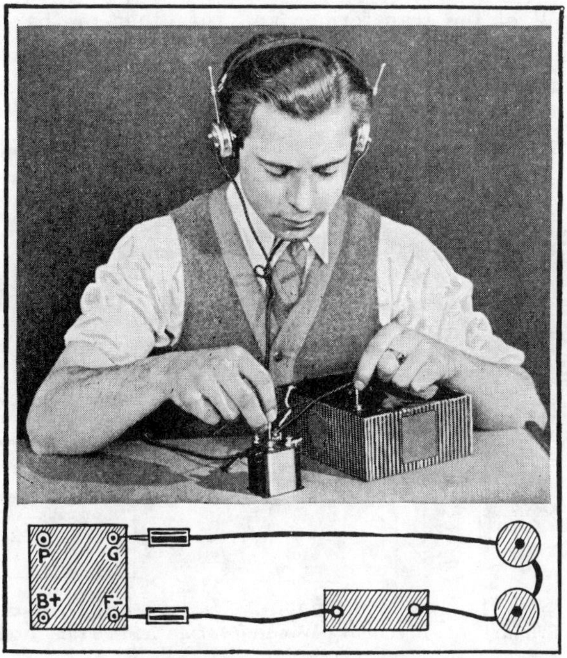

Fig. 206. Testing an Audio-Frequency Transformer for an Open Circuit with the Headphones and a B-Battery

(Courtesy of "Popular Mechanics Magazine")

when such contact is made and also when it is broken, there is leakage between the primary and the secondary windings. If a click is heard only when contact is made, the transformer is not leaky, as such a click is caused by the momentary change of charge in the windings and the accompanying rush of current through the phones.

Short circuits between the windings of a transformer, may result from contact between the internal leads to the lugs, or a grounded condition may be caused if these leads touch the transformer shell. Such trouble is usually caused by the loosening of the locknuts on the terminal screws, which permits movement of the leads. Most well-made transformers do not develop such trouble, as the internal leads are well insulated. A complete short circuit of the B-battery may result if there is a direct path for current to flow across the primary and secondary windings. To test for such a condition, touch one of the testing tips to the shell of the transformer and the other to the terminals successively. No click should be heard, and a distinct click indicates a short circuit. The trouble can usually be remedied by

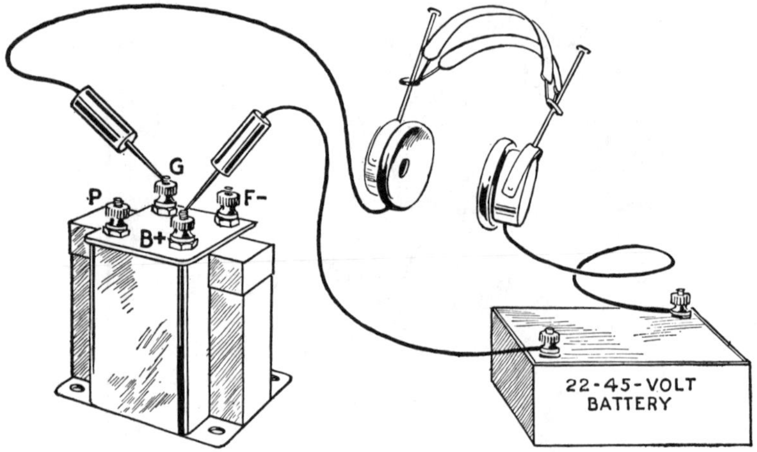

Fig. 207. Using Headphones and B-Battery to Test for a "Cross" between the Primary and Secondary Windings of an A. F. Transformer

removing the shell and providing insulation on the leads. A similar trouble is experienced when the soldering lugs, attached to the terminals on the outside, are accidentally bent down and make contact with the shell.

Radio-Frequency Transformers.—Untuned radio-frequency transformers are designed along the general lines of audio-frequency transformers, having a primary and a secondary winding, and a laminated iron core. The ratio is, however, lower than that of audio-frequency transformers, and a bakelite shell is usually used instead of a metal shell. Detail 1 of Fig. 208 shows a radio-frequency transformer. They are tested in the same way as audio-frequency transformers. The objection common to most radio-frequency transformers is that they amplify better over either the lower or the upper part of the waveband, depending on the length of the windings. Furthermore, they cannot be tuned in resonance with any signal like air-core transformers, and have a tendency to be broad. Intermediate-frequency transformers used in superheterodyne circuits, and shown in detail 2 of Fig. 208, are similar to untuned radio-frequency transformers in outward appearance.

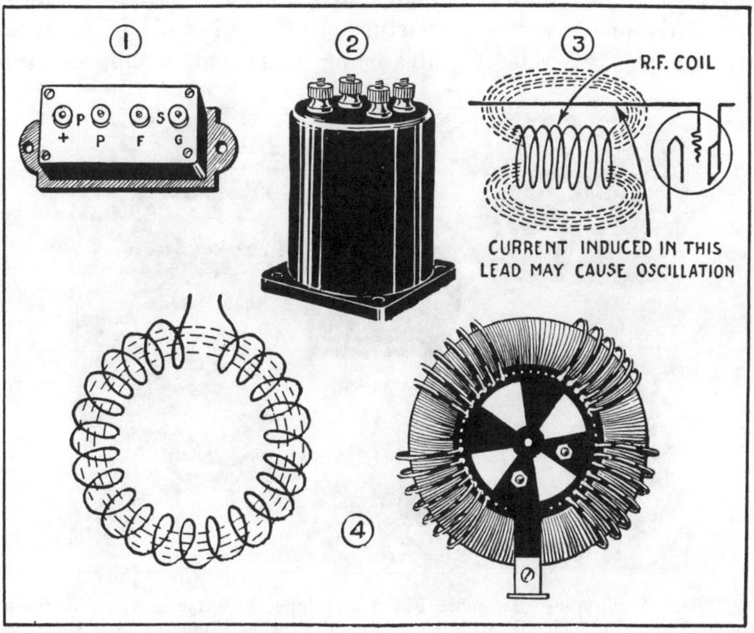

Fig. 208. (1) R. F. Transformer. (2) Intermediate-Stage Transformer of Superheterodyne. (3) How Flux Field Around Coils Induces Currents in Nearby Leads Which Can Be Overcome by Using Torroidal or "Doughnut" Coils Shown in (4)

Air-core radio-frequency transformers have been explained in the section on tuning troubles. When used in connection with two or three condensers operated by one dial, the coils must be perfectly matched so that the same settings on the condensers will tune all the coils evenly. Any kind of a coil through which a pulsating current flows sets up a field of electromagnetism, or flux, as it is commonly called. If the coil is not shielded, this field may induce currents in other leads in a receiver, as shown in detail 3 of Fig. 208. Coils of the torroidal or "dough-nut" shape do not have this effect, as their flux field is concentrated, as shown in detail 4.

Fixed-Condenser Troubles.—Good fixed condensers consist of two sets of thin plates insulated from each other by thin sheets of mica. Cheap fixed condensers often have waxed paper for insulation. To prevent moisture from getting inside and permitting losses, some fixed condensers are sealed in air-tight cases or shells. Fig. 209 shows a number of fixed condensers. Of these the center one is set in a moisture-proof shell made of moulded bakelite, the two sets of plates being connected to tapped bush-

Fig. 209. A Number of Small Fixed Condensers Varying from .0005 to 1.-mfd. Capacity. The One Shown in the Center Is Enclosed in a Bakelite-Moulded Shell Which Makes It Moisture Proof

ings to which connection can be made. Fixed condensers permit the passage of radio-frequency currents, but block audio-frequency currents, and are therefore extensively used to by-pass the former currents where they are not desired.

The troubles experienced with fixed condensers are open circuits and short circuits, usually the latter. This is caused by overloading the condenser with excessive voltage, which punctures the insulation and provides a path for current across the plates, or by damaging the insulation by the application of heat, as, for instance, from a soldering iron. For this reason the leads to be connected to a fixed condenser should first be soldered to soldering lugs, which are then attached to the condenser by means of screws, instead of first attaching them to the condenser and then soldering.

Testing Fixed Condensers.—Fig. 210 shows various steps in testing fixed condensers. They should be removed from the receiver so that there will be no shunt path around them, which would give a faulty test. The condenser is first completely discharged by holding a wire or a piece of metal across its two terminals, as shown in the upper right detail. This will prevent an erratic test caused by a residual charge. The two tips of the headphone tester, with a C-battery connected in series, are then touched to the condenser terminals as shown in the center detail. There will be a click resulting from the sudden formation of a charge on the condenser plates and the accompanying rush of current through the headphones. One of the testing tips is tapped on the same condenser terminal several times. Sharp clicks should not be heard after the first, and no click should be heard when the testing tip is removed from the terminal. A double click obtained when the tip is applied and when it is removed indicates a short circuit.

A condenser may be faulty even though it does not show a direct short circuit. Moisture in the insulation may cause a slow leakage, and this makes the condenser worthless for use in a receiver. The ability of a condenser to hold a charge can also be tested. First discharge the condenser completely and lay it on a piece of glass, mica, bakelite, or other insulating material. Charge it by holding the two leads from a C-battery or B-battery to its terminals for a moment, as shown in the lower details of Fig. 210. After waiting about two or three minutes, touch both terminals with the testing tips of the headset, which is disconnected from the battery. If a strong click is heard, the condenser has retained the charge, but if a weak click is heard, or none at all, the condenser is leaky. The strength of the click of a condenser depends to a great extent on its size. When testing a condenser, care must be taken not to touch the terminals of the condenser or the bare tester tips with the fingers, as this permits a loss through the body, and consequently results in a faulty test.

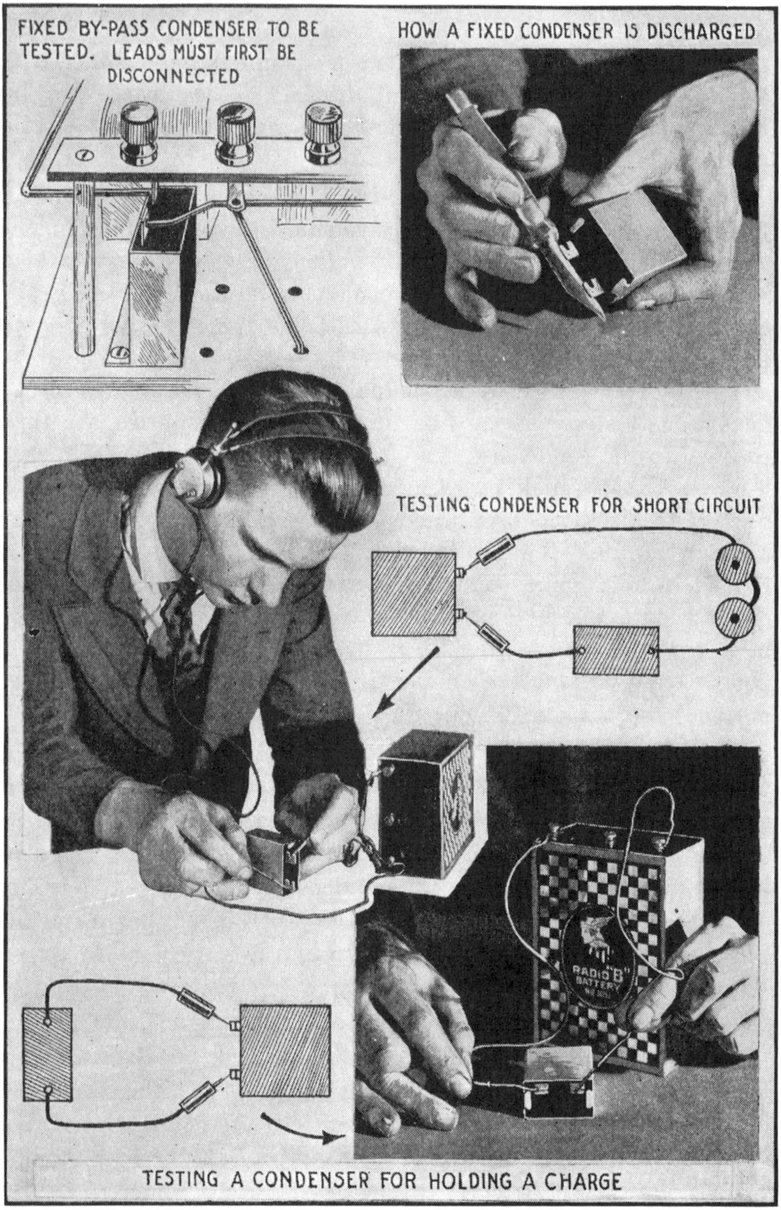

Fig. 210. Method of Testing Fixed Condensers: Above: Condenser in Receiver to Be Removed and the Method of Discharging It. Center: Testing a Condenser for a Short Circuit. Below: Testing a Condenser for Charge Dissipation

Construction of Variable Condensers.—Variable condensers using air as a dielectric are usually used for tuning purposes, being connected in the circuits in various ways. Such a condenser is shown in Fig. 211. They consist of a set of stator or stationary plates and a set of rotor or movable plates. The latter are mounted on a shaft, which extends through the panel for the attachment of a dial. Both sets of plates are well insulated from each other and are connected to terminals. Condenser plates are usually made of thin sheet brass or aluminum, being stamped out with punch presses. There are, however, some condensers that have moulded plates.

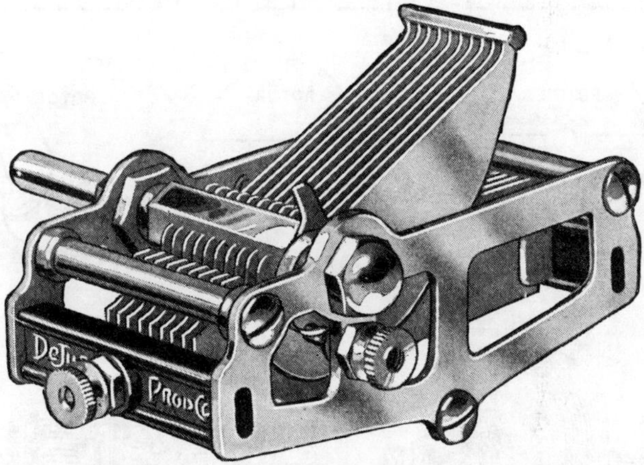

Fig. 211. Typical Variable Condenser of the Low-Loss Kind, Which Is Used Extensively in Tuned Radio-Frequency Circuits

Straight-Line Variable Condensers.—Ordinary straight-line capacity (S.L.C.) variable condensers vary in capacity in direct proportion to the dial readings, and this entails some trouble for the radio owners, as most of the broadcasting stations are then tuned in between 0 and 50 on the dials, while only a few stations come in above 50. This crowded condition on the lower dial readings results in considerable interference between stations, and makes tuning difficult, unless the receiver is ultra-selective. Straight-line-wavelength condensers (S.L.W.) spread the stations over the dials more evenly, but straight-line-frequency (S.L.F.) condensers are still better for this purpose. Details 1, 2, and 3 of Fig. 212 illustrate clearly the effect of these condensers on the dial readings. When dial readings are crowded, as just explained, the radio owner may overcome the trouble by installing S.L.F. condensers. Various kinds may be had, and it is best to take accurate measurement of the available space for the rotors to turn in, before purchasing any particular type. Details 4 and 5 of Fig. 212 show two types, the latter, as is obvious,

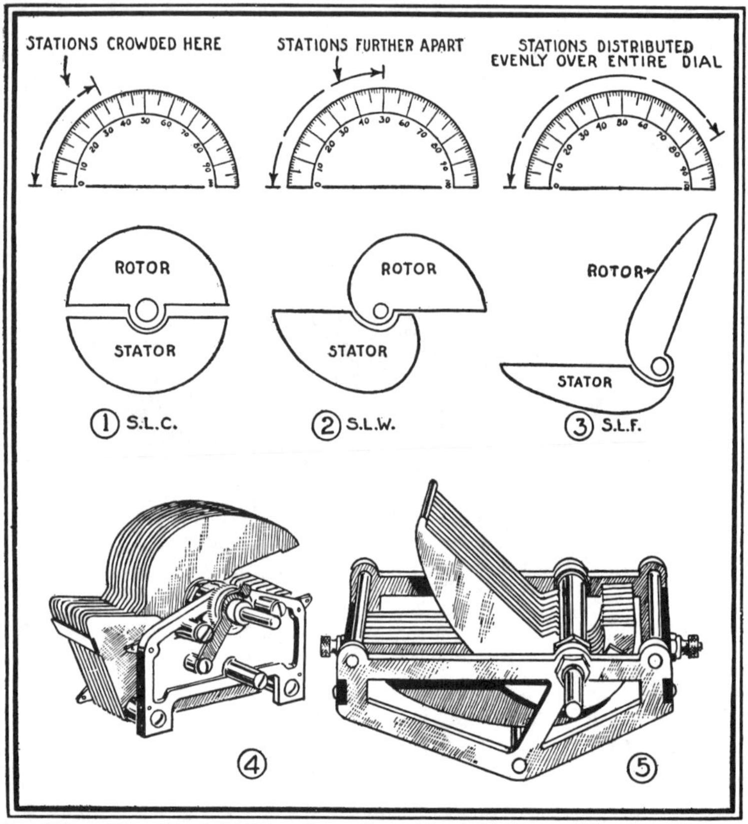

Fig. 212. (1) Straight-Line-Capacity, Straight-Line Wavelength, and Straight-Line-Frequency Condensers Showing the Shape of the Plates and the Distribution of Station Signals on the Dials. Below: Two Types of S. L. F. Condensers

requiring much more space than the former. It must be remembered, however, that S.L.F. condensers do not increase the selectivity of a receiver, but spread the stations over the entire dial more evenly, providing more space between two adjacent stations, which makes tuning easier.

Burrs on Plates of Variable Condensers.—The punch-press method of production of condenser plates entails some troubles; sometimes a slight burr is left on the edge of the plates, forming a hair-like projection, which, after the condenser is assembled, may touch an adjacent plate and permit a short circuit. Such a hair-like projection is sometimes almost invisible, but may nevertheless cause considerable trouble, which is evidenced by rasping noises in the receiver when the condenser is rotated, and sometimes by complete inaudibility of signals. To remedy this trouble take a small nail file and run it over the edges of the plates until such projections are removed.

Bent and Badly-Aligned Plates.—Condensers having thin plates are especially apt to cause trouble as the plates may be accidentally bent, causing a short circuit, or making the range of

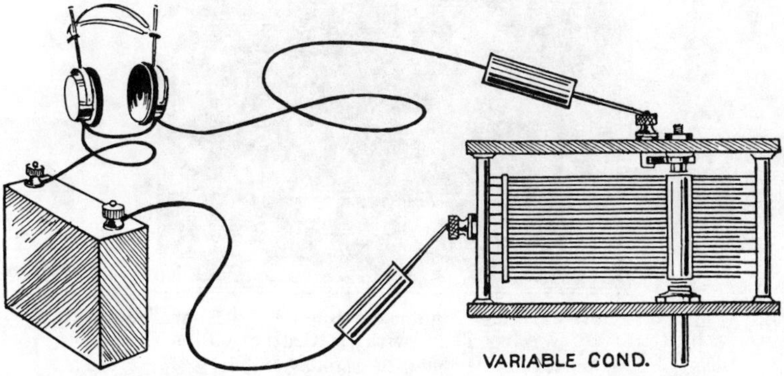

Fig. 213. Variable Condensers Should Be Tested for Short Circuits Occasionally by Means of a Headphone and C-Battery Connected in Series

capacity uneven. The distance between the stator and rotor plates should be exactly the same at any setting. This is absolutely necessary in tuned radio-frequency receivers where all the condensers are operated by one dial, uneven spacing of the plates at various settings of the dials preventing perfect balance between the condensers and the coils. Uneven spacing of the plates may also be caused by play between the shaft and the bearings, by too tight an adjustment, or by an unequal strain on the condenser, because of being improperly fastened to the panel and subpanel. The latter trouble often results when the screw holes drilled in the panel are not in perfect alignment with those in the condenser frame, and when a screw is forced down tightly the condenser frame will be bent out of shape slightly, which results in springing of the plates.

Short-Circuited and Open-Circuited Variable Condensers.—Usually one can see whether or not the plates of a variable condenser are touching each other at any point. Often this can even be heard. A test, to determine such a condition positively, can be made as shown in Fig. 213, the tips of the headset tester being attached to the two terminals of the condenser, while the dial is rotated. A click indicates a short circuit. If the whole set of rotors touches the stators on one side, the trouble can be remedied by adjusting the condenser on that side. Most variable con-

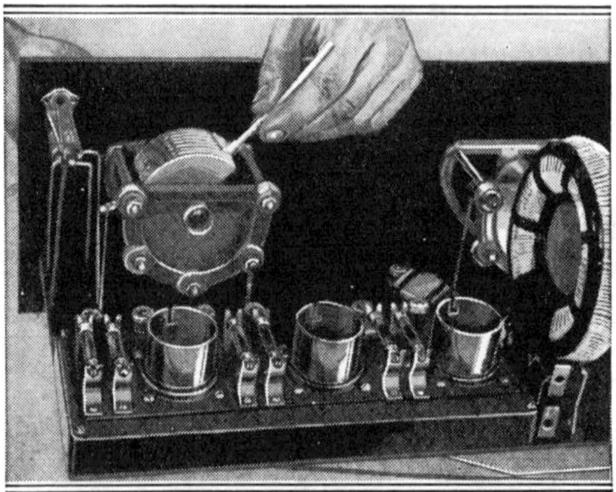

Fig. 214. Plates of Variable Condensers Should Be Kept Free from Dust by Carefully Wiping Them with an Ordinary Pipe Cleaner
(Courtesy of "Popular Mechanics Magazine")

densers are constructed so that this can be done without much difficulty. If only one rotor plate is out of line, bend it back in place carefully. To prevent the plates from being bent while transporting a receiver, it is advisable to turn the condensers so that the plates are entirely in mesh. Dust accumulates on the plates of a variable condenser and may cause a loss of efficiency. For this reason the plates should be carefully cleaned at regular intervals, which can be done by means of a pipe cleaner, as shown in Fig. 214. Many variable condensers have flexible connections between the rotor shaft and the binding post on the frame, as shown in Fig. 215. After constant turning these "pigtail" connections often become hard and brittle and then break, causing an open circuit.

Adjusting Dials of Variable Condensers.—Vernier dials are now commonly used on receivers to permit fine tuning. In many instances where ordinary plain dials are provided on receivers, the owner may remove them if he desires, and substitute new ones of the vernier type. In making this change be sure to get the right kind of dials that are suitable for the particular condenser in the receiver. Dials rotate either clockwise or counter-clockwise and are classed in this way. Most of them have graduations covering a rotation of 180°, but some have graduations covering a rotation of 360°. Be sure, therefore, to inspect the condensers before purchasing the dials. The right kind of dials to use are

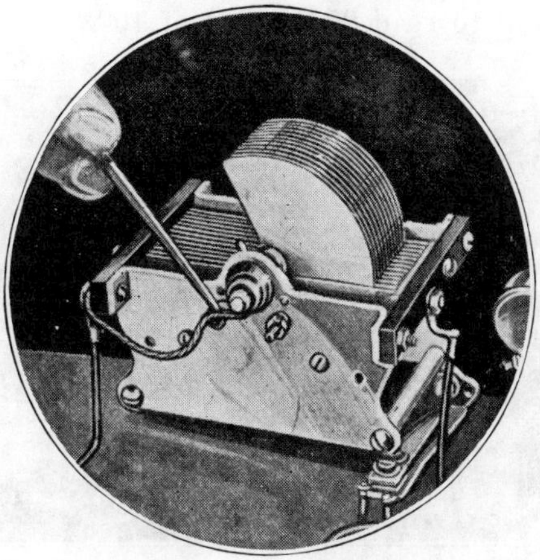

Fig. 215. "Pig-Tail" Connection to Condenser Shaft often Becomes Hard and Brittle, and Breaks

(Courtesy of "Popular Mechanics Magazine")

those that indicate a maximum reading when the condenser plates are entirely in mesh, as shown in Fig. 216.

Owing to faulty adjustment or loosening of the adjustment screw, the dial sometimes slips on the condenser shaft, which results in an erratic reading. For instance, the condenser plates may be in mesh and the pointer on the dial may indicate only 70. If this is the case, readjust the dial and condenser, and fasten the dial on the condenser shaft securely. Some dials make friction contact on the panel, and if the condenser shaft turns too hard, the dial may slip. To eliminate the trouble, adjust the condenser bearings so that the shaft can be rotated more easily.

In some cases it may be necessary to take the condenser apart and rub the section of the shaft passing through the bearing with a piece of emery cloth.

Uniformity of Dial Readings in Tuned Radio-Frequency Sets.—The dials of tuned radio-frequency receivers usually do not correspond with each other in their readings on different stations. This results from slight differences in the coils and perhaps uneven spacing of the condenser plates. If a variation of more than 5 points on the dials is observed when a station is tuned in, the dials may be adjusted so that the readings will be more uniform. For instance, if one dial is 5 points ahead of the others, reading 100, loosen it and, while the condenser plates are in mesh, set the dial to read 95 instead of 100.

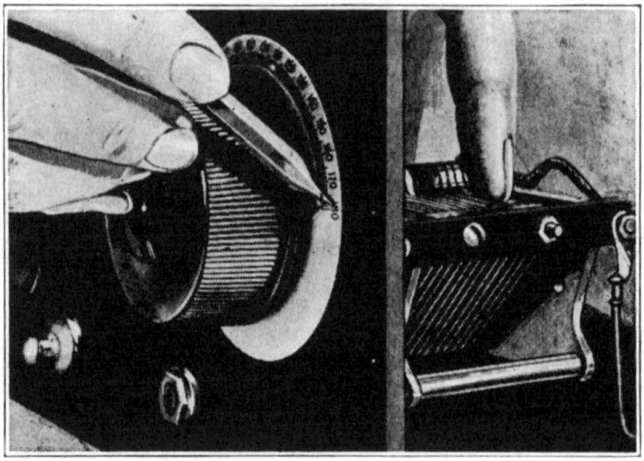

Fig. 216. When a Variable Condenser Is in Mesh Entirely the Dial Reading Should Be at Its Maximum

(Courtesy of "Popular Mechanics Magazine")

Adjusting Small Variable Condensers.—In many receivers one or more very small condensers of the variable type are often used instead of fixed condensers. If they are attached to the frame of tuning condensers and are connected to the latter, they are usually called "trimmers" or "compensators." To make adjustments on such small condensers by hand is quite a difficult task and is inaccurate, because of the capacity effect of the hand. It will be found better and more convenient to use a short length of wood or bakelite rod, as shown in detail 1 of Fig. 217, the end of which is filed down to form a screwdriver point. Never use metal for this purpose, as this will introduce capacity effects and will entail the danger of short circuits. If a small condenser has a knob and no screw slot, the knob can readily be removed, and a slot can be cut in it with a hacksaw.

Interference from Body Capacity While Tuning.—When tuning the condensers on some receivers, the strength of the signal may fluctuate with the proximity of the hands to the dials. It often occurs that the signal has been tuned in at maximum strength, and when the hands are removed from the dials, the signal decreases in volume or may be lost entirely. This trouble is caused by body capacity, and it is usually due to incorrect wiring. If

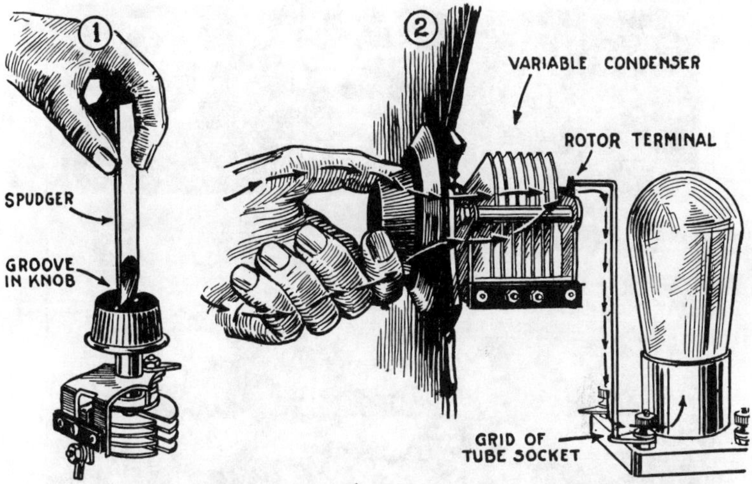

Fig. 217. (1) Adjusting Midget Condensers with a Spudger. (2) Arrows Show how Body Capacity Affects the Grid of a Tube if the Grid Is Connected to the Rotor Terminal of a Tuning Condenser

the rotor terminal of the condenser is connected to a grid, trouble may be experienced, for the rotor shaft extends through the panel, and makes contact with the dial. Then by touching the dial there will be a capacity effect, as shown by the arrows in detail 2 of Fig. 217. Reversing the connections to the condenser will often help to reduce the trouble. A grid lead to a variable condenser should always be connected to the stator plates and not to the rotor plates.

Shielding the Panel to Eliminate Body Capacity.—In some cases where body capacity is experienced and changing the connections to the condenser does not help, it may be necessary to shield the panel of the receiver with tinfoil. To do this remove the receiver from the cabinet and loosen all the instruments and brackets from the panel so that it can readily be slipped off, as indicated in Fig. 218. After the panel has been removed in this way, a piece of smooth tinfoil, which can be obtained in rolls of various lengths, is applied to the reverse side of the panel, varnish, shellac, or collodion being used as an adhesive.

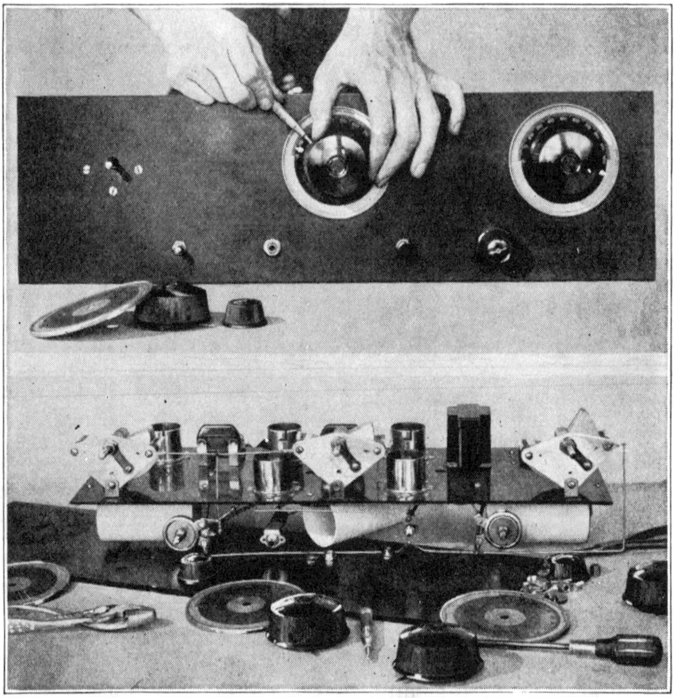

Fig. 218. A Panel Can Be Removed from a Receiver by Loosening All the Controls and Nuts Which Hold It to the Various Instruments and Brackets. There Is No Necessity of Disconnecting and Removing Any of the Instruments

After the tinfoil has been applied, hold the panel in place against the instruments so that their position can be marked accurately. With a sharp knife or razor blade, cut out space for the instruments so that when the panel is replaced the tinfoil will not touch them. It is best to clear them about ½ in. The tinfoil in the cut-out sections is removed and the adhesive is wiped off with a solvent. Use a rag saturated with turpentine to remove varnish, and alcohol to remove shellac. Ether is a solvent for collodion. After this has been done, the panel is replaced and all the instruments are fastened to it again. Such a removal and replacement of the panel can be accomplished without bending any of the wiring of the receiver, if care is taken while doing so. The shield must be grounded to the A-negative line, which can be done by letting a piece of tinfoil project to make contact with a metal bracket, which is then connected to the A-negative line, as shown in Fig. 219. Carefully examine the wiring near the panel to be sure that there is no short circuit here. If the leads near the shield are short and rigid, they can be bent away

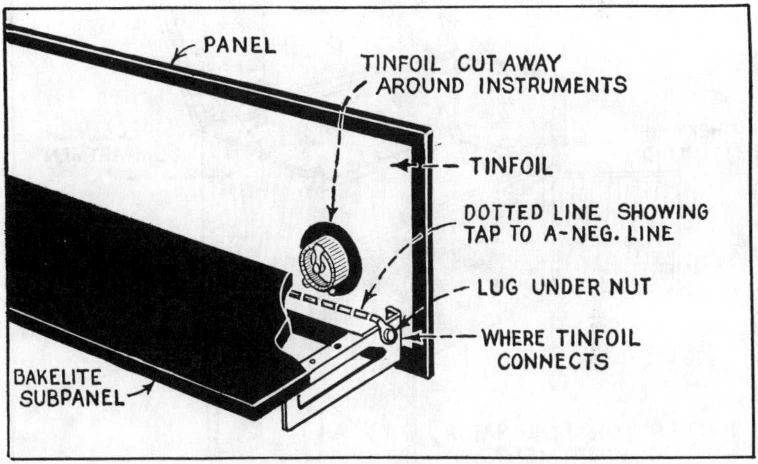

Fig. 219. Method of Using Tinfoil Shield on Receiver Panel, Showing how It Is Grounded by Connecting It to the Negative Side of the Filament Line

from the shield, but if they are long and flexible, it is better to provide "spaghetti" insulation.

Complete Shielding of the Receiver.—In order to eliminate body capacity and also any interference due to induction from sources outside of the receiver or between the individual parts of the receiver, shielding should be resorted to. One method of doing this is to apply tinfoil to the inside of the panel as described in the foregoing paragraph, and then to apply thin sheet brass, copper, or aluminum to the inside of the cabinet, as shown in Fig. 220. To do this purchase a number of sheets of light-gauge sheet brass, copper, or aluminum—never steel, iron, or the like, because of the susceptibility of these metals to electromagnetism. Then proceed to cover the entire inside of the receiver compartment with the metal, including all sides, the bottom, and the top, being sure to cover every inch of surface. The best method of doing this is to cut separate pieces for the five surfaces to be covered; tack them in place with small brass brads, and then connect all the pieces together electrically, which can be done by soldering a small wire to each of two adjoining sheets in an inconspicuous

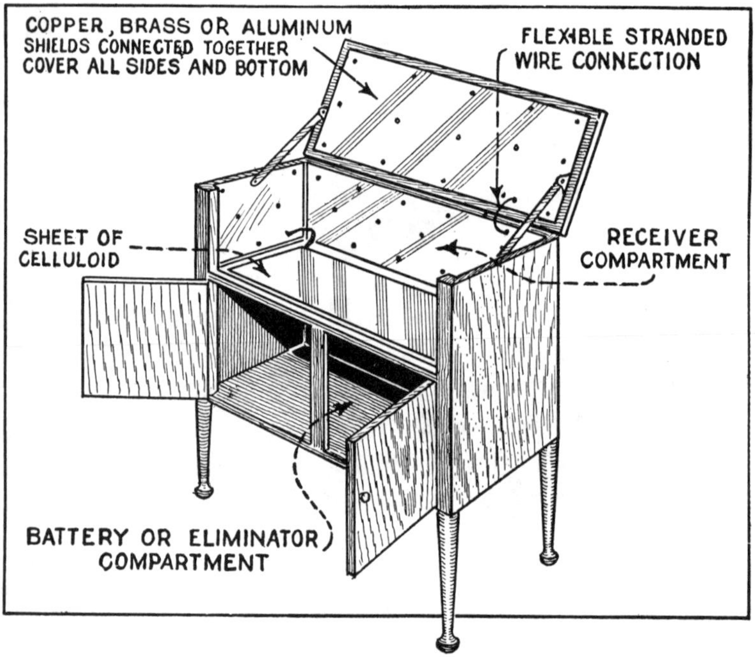

Fig. 220. Receiver Compartment of a Console Completely Shielded by Means of Copper, Brass, or Aluminum Shields Covering All the Sides, Top, and Bottom. The Panel of the Receiver Is Shielded with Tinfoil

corner. Many consoles have hinged tops and, as the top must be raised occasionally, a length of flexible, stranded copper wire should be soldered to the top sheet and to the one covering the back. Then solder a somewhat longer lead of the same wire to the bottom shield and provide a soldering lug on the free end of the lead, to make connection to the ground or A-negative terminal of the receiver. One should be careful that none of the wires or metal parts of the instruments touch the bottom sheet underneath the receiver, for this might cause a short circuit, which might burn out the tubes and injure the instruments. One method of overcoming this danger is to lay a sheet of celluloid or heavy paper over the bottom shield.

Another method of shielding is shown in Fig. 221, the individual stages being completely enclosed and isolated from each other by metal boxes, all of which are connected to the A-nega-

Fig. 221. 10-Tube Receiver Having Six Stages of Tuned Radio-Frequency Amplification, Each Stage of Which Is Completely Shielded to Prevent Mutual Inductance and Coil Pickup

(Courtesy of "Popular Mechanics Magazine")

tive line. When using this method, care must be taken to arrange the wiring so that it will not come in contact with the shield, and all holes in the shield for the passage of wires and shafts of instruments must be of ample size. "Spaghetti" insulation should be used on wires at points where they pass through the shield. The shielding boxes have removable lids in order to provide easy access to the enclosed instruments, and to permit the removal of tubes. Besides eliminating signals and inductive interference picked up by the coils and wiring of the receiver, this method of shielding prevents trouble from intercoupling between successive stages of amplification, and also makes the receiver tune more sharply.

Fixed and Variable Resistors of High Values.—Increasing use is made of high-resistance units both of the fixed and of the variable type. Resistance-coupled receivers are popular because of their tone quality. Resistance-coupling has been found highly satisfactory for audio-frequency amplification, but it is not recommended for radio-frequency amplification. With proper by-passing and balancing, resistors can be used for coupling between intermediate stages in superheterodyne receivers.

Trouble may also be experienced from faulty resistors, and therefore it is best to obtain the very best quality to avoid

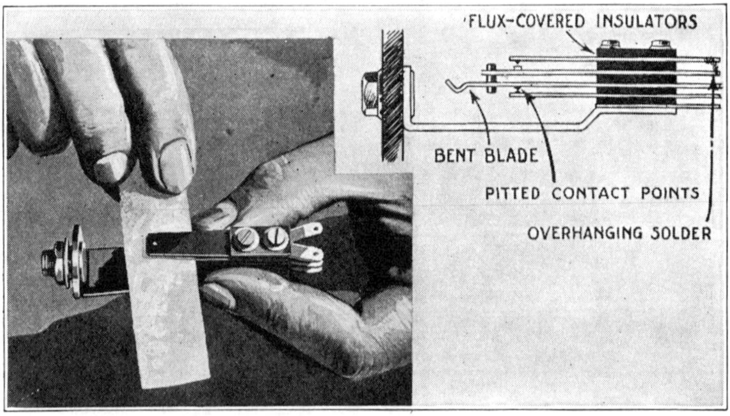

Fig. 222. Radio Troubles Can often Be Traced to Jacks Having Flux-Covered Insulators, Bent Blades, Pitted Contacts, and Short Circuits. Sandpaper Is Useful for Cleaning the Contact Points

trouble. Most fixed resistors consist of a glass tube with metal end caps with the resistance element placed inside and securely connected to the end caps. The connection may be broken and the resistor is then useless. Many resistors are inclined to be noisy owing to changes in the resistance element, which is affected by atmospheric and temperature conditions. Some manufacturers produce resistors having the resistance element impregnated in the glass. Resistors made this way are claimed to be superior to the common type, as they are more stable in their operation. It is a good idea to keep an extra set of fixed resistors on hand so that they can be substituted for those in the receiver in case the latter are suspected to be defective. When replacing resistors be sure to get the same value as that of the defective one, or trouble may result.

Jacks and Their Troubles.—Fig. 222 shows a drawing of a filament-control type of jack and the points where trouble is liable to develop. Sometimes the spring blades become bent and make faulty contact with the plug terminals or with the contact points of other blades. The contact points may also become pitted and burned, owing to arcing. This trouble can often be remedied by cleaning the contact points with a nail file or sandpaper, as shown in the photo, Fig. 222. A film of corrosion on the blade that makes contact with the plug will also create a high

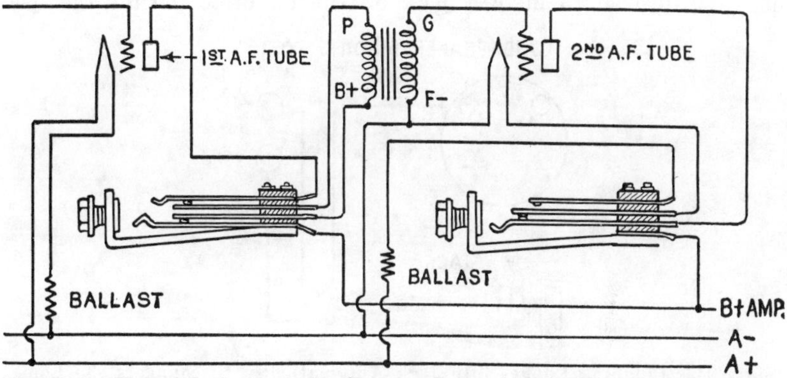

Fig. 223. Use of Jacks in Audio Stages. Removing the Plug from the Last Jack Disconnects the Filament of the Last Tube and Inserting the Plug into the Second Jack Cuts out the Primary of the Last Transformer

resistance and reduce the efficiency of the receiver. Soldering flux sometimes runs over and between the insulating shims of a jack and permits leakage. In some cases arcing may even occur over flux left between adjacent blades. Flux should therefore be removed by means of alcohol, a small brush being handy for this purpose. Short circuits may exist between the ends of the jack blades at the points where the wires are connected, due to drops of solder hanging down from one terminal to the next. This trouble can readily be remedied by merely resoldering and removing some of the excess solder.

Single-circuit jacks are used to conduct the plate current from the last audio tube or output transformer to the loudspeaker.

Filament-control jacks have the additional feature of serving as a switch in the filament line to light the tubes when the loudspeaker is plugged in. Two jacks are sometimes provided on receivers, one supplying the output of the first audio tube and the other the full output of the receiver. The former is used for headphone reception of long-distance stations or loudspeaker reception of nearby stations, enough volume being obtained without the use of the last tube. In such a double-jack arrangement the filament of the last tube is automatically disconnected when the plug is removed from the last jack. A wiring diagram of this arrangement is shown in Fig. 223.

When receivers are installed in consoles, the loudspeaker connections are usually made inside of the set. In such cases it may be desired to have a jack on the panel to permit the use

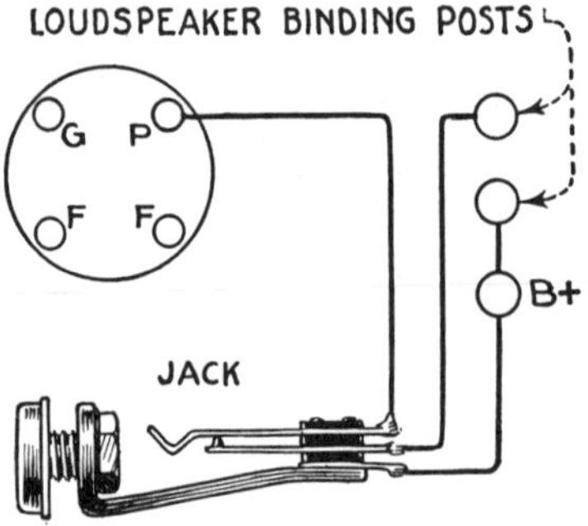

Fig. 224. Plugging Phones into Jack Automatically Disconnects the Loudspeaker and Vice Versa

of headphones for distant reception, and then the wiring arrangement shown in Fig. 224 is followed. With this arrangement the loudspeaker can be left connected permanently, but electrical contact with it is broken automatically when the phone plug is inserted into the jack. Removing the plug again puts the loudspeaker in the circuit.

Replacing a Broken Panel.—If a panel has been broken it can be replaced without much difficulty, provided the proper tools are used for the purpose. Remove the broken panel in the same way as explained in the paragraph on "Shielding the Panel to Eliminate Body Capacity." Then place the broken panel on the new one, which should be the same size, and hold the two panels securely together with small C-clamps. With a scriber, trace the holes of the broken panel on the new panel. Small screw holes can be drilled by using the old panel as a template, and the center of the large holes, which require reaming, can be located by means of a prick punch. The drilling is done with a hand drill, as shown in Fig. 225. C-clamps may be used to hold it securely on a piece of wood if necessary. A small vise to hold

Fig. 225. Replacing a Broken Panel Necessitates Careful Duplication of Holes in the New Panel with the Hand Drill
(Courtesy of "Popular Mechanics Magazine")

the panel is still better. If the drills are not large enough, make the holes larger by means of a reamer. A countersink is used in the drill for countersinking holes for flat-head screws. A satin finish can be had on the panel by rubbing off the gloss with fine sandpaper. When doing this be sure to rub in one direction only. If indicator marks are needed on the panel over the tuning dials, these can be made by means of the scriber, drawn along the edge of a rule, and the groove thus made can be whitened by rubbing ordinary chalk into it.

Operation and Care of Crystal Receivers

A crystal receiver usually consists of a variable inductance as a variometer, vario-coupler, or tuning coil of some kind; a variable condenser, a crystal detector, and a pair of headphones, connected as shown in Fig. 226. No batteries are required for its operation and it gives remarkably good tone quality, but only enough volume for headphone reception. The maximum range of the average crystal receiver is approximately 15 miles. For best results the aerial of a crystal set should be from 150 to 200

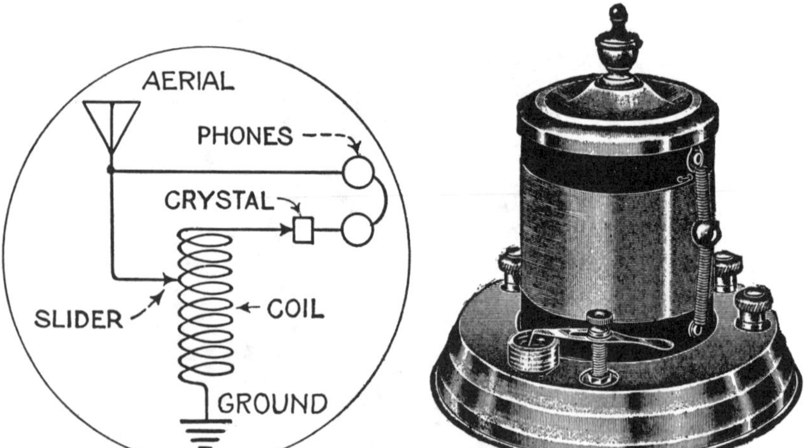

Fig. 226. Left: Diagram of a Simple Crystal Receiver Using a Tuning Coil Having a Variable Inductance. Right: Commercial Form of an Inexpensive Crystal Set Made in a Compact Unit

ft. in length, so that there will be a maximum pick up. A good ground is also essential. The crystal used to detect the signals should never be touched with the fingers, for a film of oil is then deposited on the surface and this tends to reduce its sensitivity. Occasionally a crystal should be cleaned by brushing it with alcohol and if this does not help to renew its sensitivity it should be replaced with a new one. It will be found that one crystal may be much more sensitive than another. Usually a "cat-whisker" contact is made with the surface of the crystal, and it is necessary to maintain this contact for good reception. The holder of the "cat-whisker" should be of a type which can be tightened. Fixed crystals can also be obtained, and this type is preferred by many, as it does not have to be adjusted. Crystal sets, which are exposed, should be kept free from dust. Broken connections in the wiring may result from jarring, but as a rule such breaks can easily be found. Frequent handling of the head-set may result in breaking the flexible cords, which results in complete inaudibility. New cords cost little. Avoid jarring the phones by setting them down carelessly, as violent vibration tends to decrease the strength of the small magnets inside.

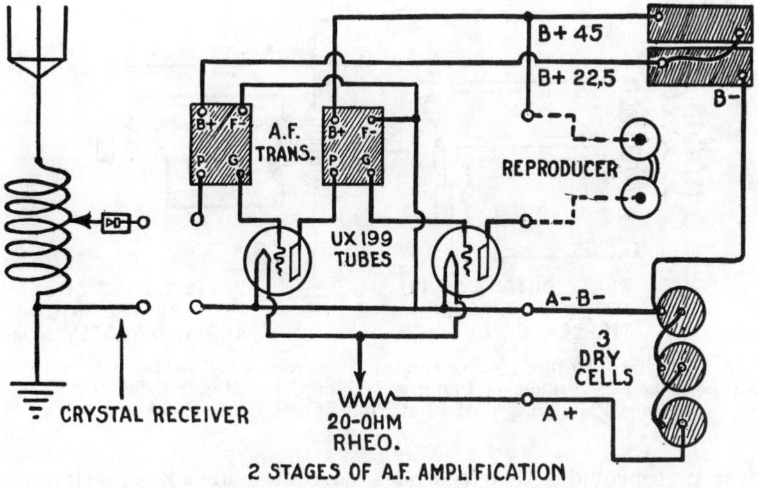

Fig. 227. Two-Stage Audio Amplifier, Which when Connected to a Simple Crystal Set, Greatly Increases Its Range and Volume. Dry Cells and Two 22.5-Volt B-Batteries Are Used

How to Add a Two-Stage Amplifier to a Crystal Set.—If the owner of a crystal set desires loudspeaker reception of local stations, a two-stage amplifying unit can be added to a crystal detector, as shown in Fig. 227. It consists of two $3\frac{1}{2}$ to 1-ratio audio-frequency transformers, two UX 199 tubes, two tube sockets, a 20-ohm rheostat, and nine binding posts. The unit can be built in a small radio cabinet if desired. The output of the crystal is connected to the input terminals of the amplifier unit, and the output of the latter is connected to a loudspeaker. The amplifier needs 6 dry cells and a 45-volt B-battery for its operation. All the wiring details and the proper method of hooking up the batteries are shown in the diagram. All connections should be securely soldered. Either No. 14 round copper busbar wire or flexible wire can be used.

Controlling Reradiation of Regenerative Receivers

Any type of set, which feeds energy from the plate output of one or more tubes back into the aerial circuit directly or indirectly, is a regenerative set. Usually the plate output of the

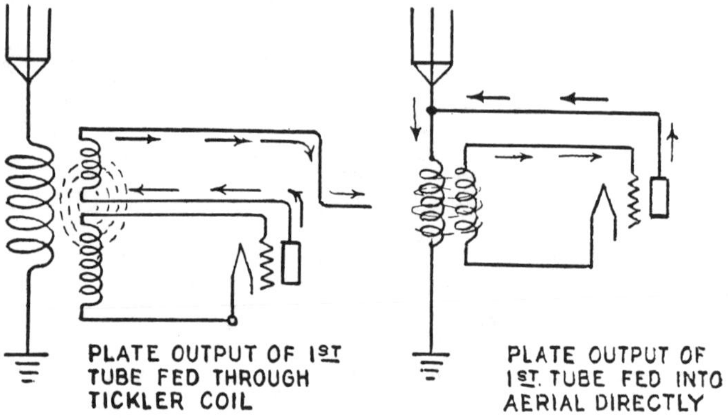

Fig. 228. Two Methods of Obtaining Regeneration. Left: Regeneration Accomplished by Induction Using a Tickler Coil. Right: Feeding Part of the Plate Output of First Tube Directly into the Aerial

first tube provides this feed-back current, which is superimposed by induction on the tuning coil indirectly by means of a tickler coil, as shown in the left detail of Fig. 228, or the plate output is directly connected to the aerial circuit as shown in the right detail of Fig. 228. When the feed-back energy is too strong to be absorbed by the first vacuum tube, it begins to oscillate and the aerial then reradiates signals into the air, which can be heard in other receivers several miles away. Such reradiation seriously interferes with the reception on other receivers in the neighborhood, and it is often evidenced by "mocking-bird" whistles commonly called "bloopers."

Non-Reradiating Unit.—Radio owners who are troubled by reradiation from neighboring regenerative receivers cannot eliminate the trouble, but the owner of the offending receiver can prevent it by installing a non-reradiating unit, which will not interfere with his own reception, and will eliminate the objectionable whistles in his neighbors' reception. The diagram of such a unit is shown in Fig. 229. It consists of four small coils made of No. 26 insulated wire. Coil 1 is wound on a piece of 2¾-in. tubing about 3 in. long; it has 65 turns of wire, a tap being taken at the 15th turn as indicated. Coils 2 and 3 are both wound on one tube, which is 3 in. in diameter and 4 in. long. By taking two wires in hand, both coils can be wound at the same time. The end turn of one coil is connected to the first turn of the other. Coil 4 is wound on the end of the same tube,

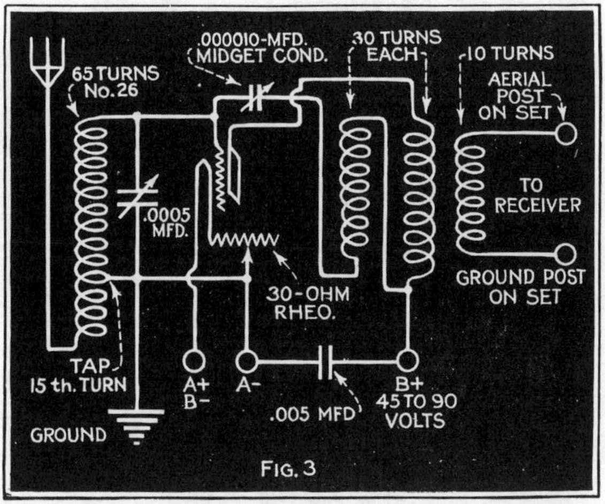

Fig. 229. Diagram of a Specially Constructed Unit Which Prevents Reradiation of a Regenerative Receiver, Providing an Additional Stage of R. F. Amplification and also Serving as a Wave Trap
(Courtesy of "Popular Mechanics Magazine")

and consists of 10 turns. Two variable condensers, one of .0005-mfd. capacity, and the second of .00001-mfd. capacity, or if this size is not obtainable, one of 2 to 35-mmf. capacity, a tube socket, and a 30-ohm rheostat for a UX 199 tube, or a 6-ohm rheostat for a UX 201A tube, are provided, and the coils and instruments are hooked up as shown in the diagram. A separate cabinet should be provided for this unit. When completed, the unit is connected to the aerial, ground, receiver-input terminals, and the batteries.

It is then necessary to balance the unit, which is done in the following way: Turn on both the receiver and the unit, and set the tuning condenser on the latter at 40. Turn up the receiver until it oscillates, and tune in the signals from a broadcasting station, which will come in with a whistle. Bring in the signal to the point of maximum volume by adjusting the tuning dial of the unit. Then turn the small condenser to a point where, upon turning the large condenser on the unit, the whistle will not vary in pitch but only in volume. The small condenser should be kept at this point permanently as the unit is then balanced. Tuning is accomplished by the receiver dials and also by the condenser on the unit.

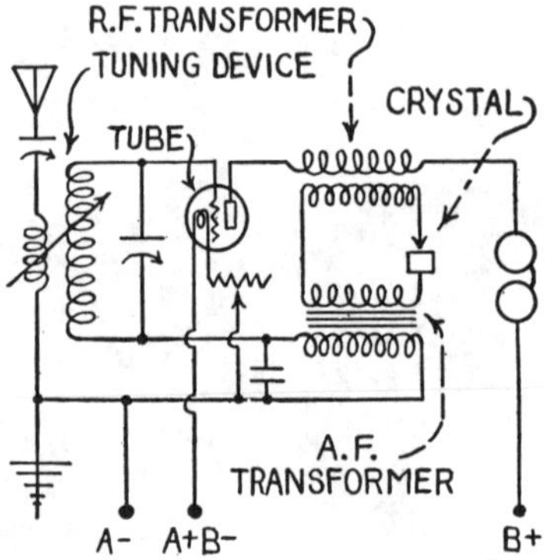

Fig. 230. Diagram of a One-Tube Reflex Receiver Using a Crystal for a Detector

Reflex Receivers

In reflex circuits some vacuum tubes act both as radio-frequency and as audio-frequency amplifiers. In Fig. 230 is shown a one-tube reflex receiver having one stage of radio-frequency, a crystal detector, and one stage of audio-frequency amplification. As a rule a reflex set has three controls: a coupling control, a condenser for turning the aerial, and a grid-control condenser. If the coupling control is left unaltered, a station will always come in on the same dial readings. When tuning a reflex set, rotate the grid condenser until a station is brought in. The volume can be increased by adjusting the aerial-tuning condenser. If there is any interference, loosen the coupling and readjust the tuning condenser.

The most frequent trouble experienced with reflex receivers is oscillation. Sometimes this can be eliminated by reversing the connections between the radio-frequency and the audio-frequency transformers. Proper adjustment of the filament rheostats will be found to prevent oscillation. Reflex sets may employ either a crystal or a vacuum-tube detector; those having a crystal detector have the better tone quality, as the noise and distortion of an extra tube is eliminated, but vacuum-tube detectors are often used. It is advisable to use the large 5-volt tubes in reflex circuits as the mutual conductance or plate-current flow of the 3-volt tubes is so small that they do not function as effi-

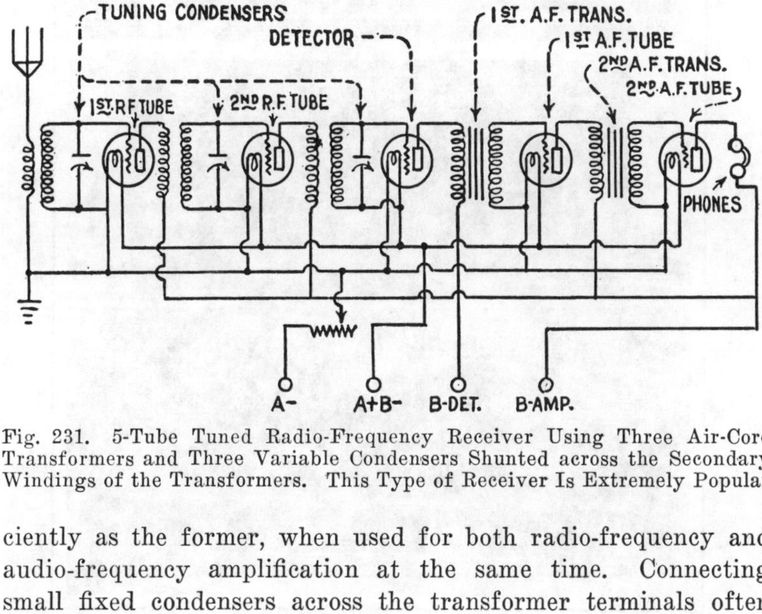

Fig. 231. 5-Tube Tuned Radio-Frequency Receiver Using Three Air-Core Transformers and Three Variable Condensers Shunted across the Secondary Windings of the Transformers. This Type of Receiver Is Extremely Popular

ciently as the former, when used for both radio-frequency and audio-frequency amplification at the same time. Connecting small fixed condensers across the transformer terminals often helps to reduce oscillation by by-passing the radio-frequency currents.

Tuned Radio-Frequency and Neutrodyne Receivers

Tuned radio-frequency and neutrodyne receivers are practically the same in operation except that the internal capacity of the radio-frequency stages is neutralized in neutrodyne receivers. Most sets of this kind have an aerial-tuning coil and condenser, and a similar tuning coil and condenser for each of the two radio-frequency stages. Three or more stages of radio-frequency amplification are seldom used. The schematic wiring diagram of a simple tuned radio-frequency set is shown in Fig. 231, and

Fig. 232. 6-Tube T. R. F. Receiver Having Transformer and Resistance-Coupled Audio Stages. The Coils and Wiring Are Concealed under the Subpanel

three views of a receiver of this type are shown in Fig. 232. Regeneration can also be used in these circuits, but this does not alter their tuning arrangement. The three coils are usually fixed, but the three condensers are of the variable type. These may be mounted on the subpanel separately, and are then operated by three different dials, or they may be incorporated in one tandem condenser of the kind shown in Fig. 233.

To tune receivers of this kind, turn on the filament current to light the tubes and advance the rheostats controlling the audio and radio-frequency tubes about two-thirds. Then advance the tuning dials slowly, keeping them practically at the same dial readings. When a station whistle is heard, tune in the whistle on each dial separately, and when the point of maximum volume is reached it may be necessary to turn down the rheostats. This is usually the case when the receiver is in close proximity to the broadcasting station and signals are brought in too loud for comfortable reception. After the rheostats have been turned down in this way, the signal should again be tuned in at the peak of the wave by adjusting the dials.

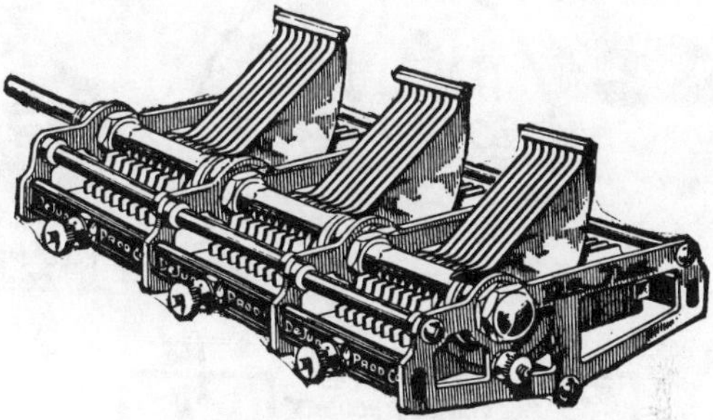

Fig. 233. Three Condensers Mounted on One Shaft Forming a Tandem Condenser Which Is Used on Many One-Dial Receivers

How to Neutralize a Neutrodyne

Oscillation in neutrodyne receivers arising from the internal capacity of radio-frequency tubes is eliminated by means of small balancing condensers or "neutrodons," which are connected between the plate and grid elements of successive tubes, as shown in Fig. 234, the grid connection being made to a tap in the secondary winding of the transformer, as indicated. These small condensers must be adjusted before the receiver will function properly as a neutrodyne, and the process of making these adjustments is called neutralizing. To do this it is necessary to obtain a strong signal from a broadcasting station, which is usually possible in large cities where several stations are located. If, however, the receiver is located a considerable distance from a station, and no extremely loud signals are possible, then a buzzer-operated oscillator or wavemeter, producing radio-frequency oscillations, must be used.

Such a device is shown in Fig. 235, and it can readily be made by anyone. It consists of an ordinary door-bell buzzer, a .0005-mfd. variable condenser, a single-pole knife switch, a 70-turn inductance coil, wound on a 3-in. bakelite or cardboard tube, and two dry cells. By rotating the condenser various wavelengths can be obtained. In use it is located about 20 ft. away from the receiver to be adjusted, and it is connected to the aerial post as shown, the ground post of the receiver being connected to the ground in the usual way. The switch on the buzzer unit is closed, the buzzer is set so that it gives a clear high-pitched

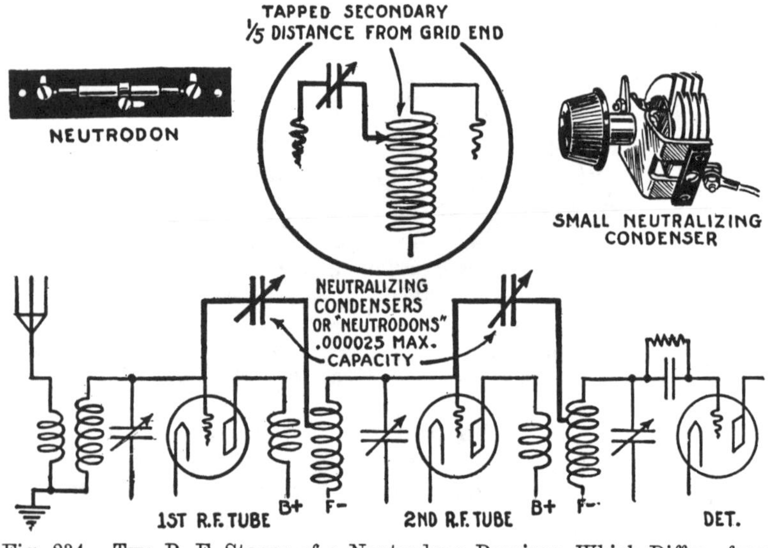

Fig. 234. Two R. F. Stages of a Neutrodyne Receiver, Which Differs from an Ordinary T. R. F. Receiver in That Small Neutralizing Condensers Are Used to Balance the Internal Capacity of the Tubes

sound, and the condenser dial is then set about 10 degrees, which will cause the buzzer to oscillate at a wavelength of approximately 225 meters. The dials of the receiver are rotated to a point where the signal comes in loudest.

If no buzzer is used, the same method of procedure is followed by tuning in a 225-meter station. After the signals have been tuned in as loud as possible, remove the first radio-frequency tube from its socket. The signal will become fainter, but the tuning dials are again carefully readjusted to bring in the signal as loud as possible with this tube removed. The first radio-frequency tube is the one closest to the tuning unit. Slip a piece of paper over one of the filament tips of the re-

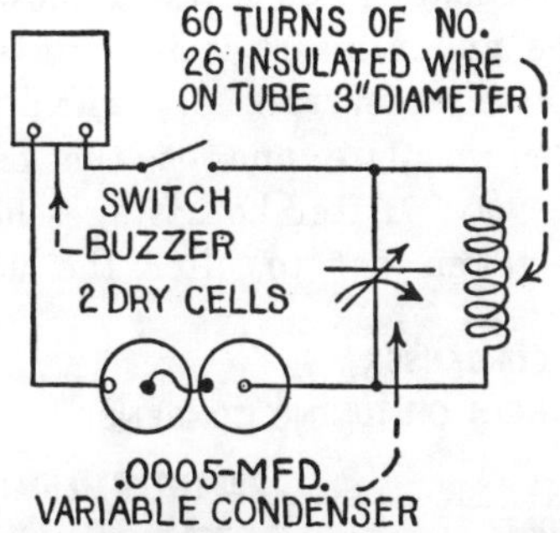

Fig. 235. Buzzer-Excited Oscillator Used for Neutralizing Neutrodyne Receivers and Testing Superheterodynes

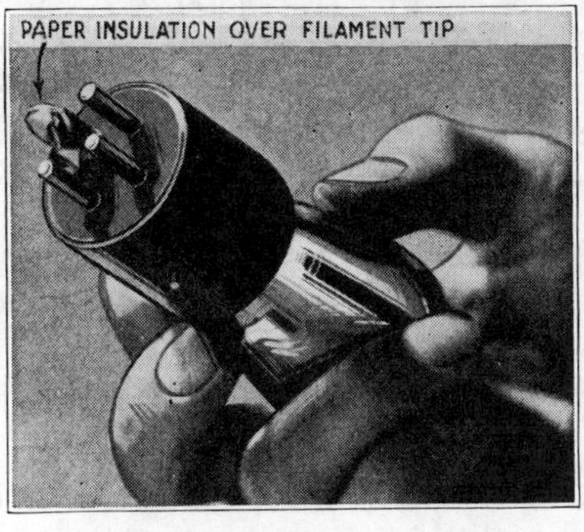

Fig. 236. Method of Insulating the Filament Tip of a Tube When Neutralizing a Neutrodyne Receiver

moved tube, as shown in Fig. 236. The filament prongs are the heavy ones, in case of a UX 201A, CX 301A, or similar type of tube. If the tube is one of the UV 201A or C 301A type, not having heavy tips, the two filament tips are those farthest from the pin on the side of the base.

With one filament tip insulated in this way, insert the tube into the socket. It will not light up. The tuning dials should not be touched, but the small neutralizing condenser between the first and second stage is adjusted to a point where the signal is nearly or entirely inaudible, and it is then left at this point. Adjustment of this condenser should be made with a spudger. Do not attempt to make this adjustment by hand or with a metal tool, as body capacity and capacity of the tool will interfere. After the adjustment is made, remove the paper insulation from the tube and reinsert it into its socket. The first stage is now completely neutralized. Then remove the next radio-frequency tube, insulate one of the filament tips, and proceed in the same way. After both stages have been neutralized, care must be taken not to alter the adjustment of the

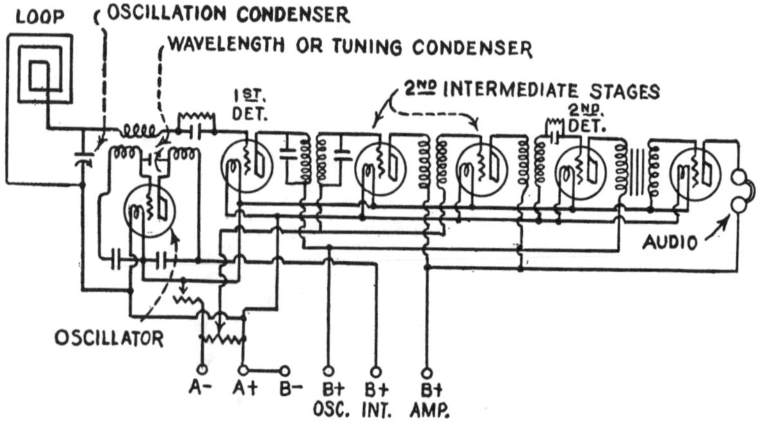

Fig. 237. Schematic Wiring Diagram of a Typical 6-Tube Superheterodyne Receiver, Showing Oscillator, First Detector, Two Intermediate Stages, Second Detector, and One Audio-Frequency Stage

small condensers. Sometimes they can be screwed down securely so that there will be little or no danger of their being changed. When it is necessary to replace the radio-frequency tubes, be sure to substitute tubes of the same type, which have the same characteristics. If different tubes are used, the neutralizing process will have to be repeated.

Superheterodyne Receivers and Their Troubles

If properly constructed, a superheterodyne receiver is extremely selective and is capable of tuning in outside stations through locals. Because of the selectivity of this type of receiver outside stations can usually be tuned in without interference, but its range is slightly greater than that of a 6-tube tuned radio-frequency receiver. Improper tuning of superheterodyne receivers often results in considerable noise, and if the receiver is incorrectly designed and poorly constructed, it will also cause trouble. If it is of the home-made variety, care should be taken to arrange the instruments and wiring exactly as specified in the blueprints and instructions.

A superheterodyne consists of a first detector, an oscillator, a number of intermediate-frequency stages, a filter unit, a second detector, and one or more stages of audio-frequency amplification. The diagram of a superheterodyne receiver is shown in Fig. 237. When operating properly there should be a sound in

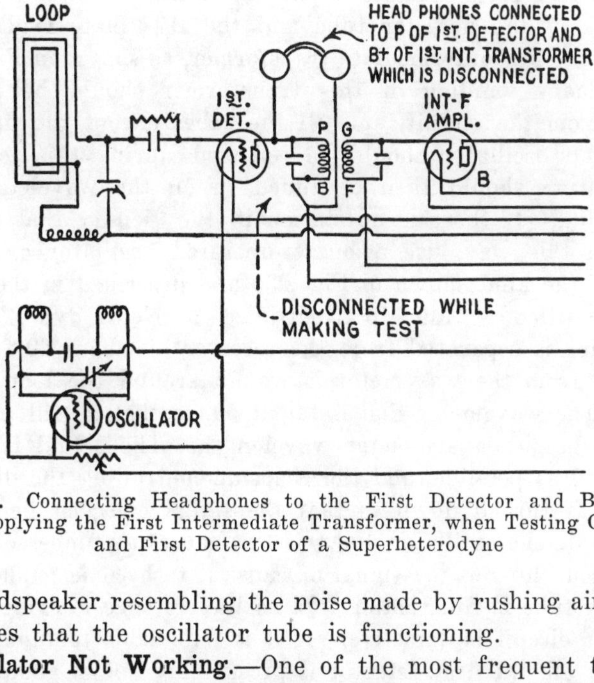

Fig. 238. Connecting Headphones to the First Detector and B-Positive Line Supplying the First Intermediate Transformer, when Testing Oscillator and First Detector of a Superheterodyne

the loudspeaker resembling the noise made by rushing air. This indicates that the oscillator tube is functioning.

Oscillator Not Working.—One of the most frequent troubles experienced with superheterodyne receivers is that the oscillator does not work. Some tubes will serve better than others as oscillators, and therefore all the tubes in the receiver should be tried to find which is the best one for this purpose. When the oscillator does not function, the "rushing-air" sound will no longer be heard, and the receiver seems to have lost some of its energy. Usually stations can still be tuned in, even though the oscillator is not functioning, but then the receiver is working merely as a radio-frequency receiver. In such cases signals can be tuned in by rotating the wavelength or tuning dial only, and rotation of the oscillator dial will not affect tuning at all.

To test whether or not the oscillator tube is oscillating, touch the grid terminal of its socket with a moistened finger or with a metal rod. This should cause a click in the loudspeaker when the tube is oscillating. Another simple method of testing the oscillator is to tune in a station, if possible, and then remove the oscillator from its socket; if the signal continues to come in, it is proof that the oscillator tube is not functioning. A still more accurate test is to connect a pair of headphones to the plate terminals of the detector tube and the B+ plate lead which supplies the first intermediate transformer, as shown in Fig. 238. The primary winding of this transformer should be disconnected from the circuit, and all the tubes except the first detector and oscillator should be removed during this test. A local station should then be tuned in on the wavelength or tuning dial. If there is no station in the vicinity that can be heard in this way, use a buzzer-operated oscillator or wavemeter of the kind shown in Fig. 235 and described in the paragraph entitled, "How to Neutralize a Neutrodyne." The wavemeter is separated from the receiver by about 20 ft., and the lead from the wavemeter is wound around the loop a few times. The wavemeter dial is set at 50, so that it will produce a signal having a 400-meter wavelength. The signal is tuned in as loud as possible, and the rheostat controlling the detector filament is turned down so that the signal can just be heard. Then rotate the oscillator dial. If this causes a number of whistles, which blur out the signal or cause it to become muffled, the oscillator tube is functioning. If no interference is evident, the oscillator circuit is defective. If it is impossible to tune in the signal at all, the first detector is presumably not in good condition, or there may be a break in the wiring.

The first remedy which may be found to eliminate the trouble is to use another tube in the oscillator socket, as the trouble may be due to a poor tube. Examine the connections on the coupling unit and tubes, to make sure that there is no open circuit. Perhaps the plate and grid leads to the coupling unit are reversed. The plate lead must always be connected to the outside winding, and the grid lead to the inside winding of the coupler. If any section of the coupling coil, a condenser, or the primary winding of the first intermediate transformer is short-circuited, the oscillator will not function. Too little or too much plate voltage on the oscillator tube will also prevent it from working. Sometimes lack of oscillation is caused by too high capacity of the midget feed-back condenser in the plate of the first detector. This condenser should have a minimum capacity of .00005-mfd. (or 15 micro-microfarads, mmfd.). When a new oscillator tube is used, this condenser must be adjusted accurately, which is done by turning it so that all the plates will be out of mesh. After a station is tuned in the capacity of the feed-back condenser is increased so that signals will come in at the maximum volume without permitting the detector to oscillate.

Troubles of the Intermediate-Frequency Amplifier.—The intermediate-frequency transformers must be matched in order to obtain maximum efficiency from superheterodyne receivers. Poor selectivity is often caused by unmatched transformers. Sometimes the amplification of the intermediate-frequency stages drops as low as 50 per cent below normal, because of unmatched transformers. The transformers should not be set too close together, for in some types of transformers this may result in coupling effects, which will cause howls and squeals. It is best to keep the intermediate transformers at least 2 in. apart, especially if they are of the air-core type. In superheterodyne circuits not using a stabilizer or "losser" with the intermediate transformers, there will be a tendency to oscillate when the filaments are at their normal operating point. Intermediate transformers are designed to operate at a definite frequency, which varies in different types of transformers. A transformer designed to operate on a high frequency will be more selective than one designed to operate on a low frequency. The objection to the former type is that it tunes in only the peak of the wave, and causes a loss of tone quality. The latter type has a tendency to tune broad, but this is compensated for by the improvement in tone. Squealing, when the volume-control rheostat is advanced, is sometimes caused by a feed-back from the audio-frequency stages. This trouble can usually be remedied by grounding the metal shell of the audio-frequency transformers to the negative filament line. A radio-frequency choke coil, shunted with a .002-mfd. fixed condenser and connected in the plate line of the second detector, will also help. Superheterodyne receivers having resistance or impedance-coupled audio amplification are usually difficult to control. If the loop leads or the wiring from the loop binding posts to the wavelength or tuning condenser are too close to the intermediate transformers, there may be considerable loss of selectivity. This is due to the pick-up value of the transformers.

The Filter Unit.—Every superheterodyne has a special transformer, which is called the filter. It is shunted by condensers of definite capacities, and it is tuned so that nothing but the desired waveband will be passed. In some circuits the filter transformer is placed ahead of the intermediate transformers, and in other circuits it is placed just ahead of the second detector, depending on the design of the instruments and the circuit. Loss in selectivity may be caused by a poor or defective filter, or by a filter condenser of incorrect capacity. Whistling and howling sounds or total inaudibility may also be caused by an incorrect value here.

Tuning a Superheterodyne Receiver.—When tuning a superheterodyne receiver, point the loop in the direction of the station to be tuned in. Then adjust the wavelength or tuning condenser to the point where signals at this wavelength should come in on the dial. This point can be approximated by assuming that a station operating on a wavelength of 400 meters, which is halfway between 200 and 600 meters (the broadcasting waveband), will be tuned in about halfway on the dial. The dial setting for other stations can be calculated roughly in the same way. After the tuning dial has been set, the oscillator dial is moved very slowly over a range of about 10 points on each side of the setting of the tuning condenser. While moving it, a short whistle may be heard, which is the station whistle. Keep the oscillator dial on the whistle and then readjust the wavelength dial very slowly until the signal comes in. Both dials are carefully rotated to the maximum point of volume. It should always be remembered, when tuning receivers, especially superheterodyne receivers, that the tuning dials must always be rotated very slowly. Distance reception depends especially on painstaking tuning and patience on the part of the operator. Selectivity can often be increased by turning the rotor in the coupling coil so that the primary and secondary windings will not be in the same plane; this loosens the coupling and reduces the inductive relation between the coils. If this is done, it may be found necessary to readjust the midget feed-back condenser in the plate circuit of the first detector.

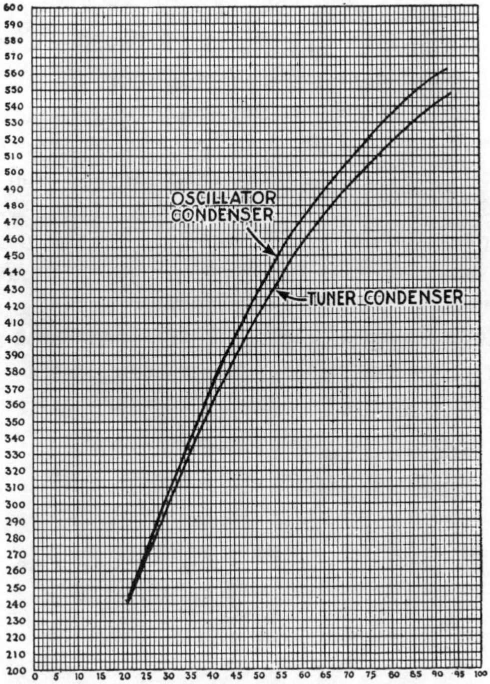

Fig. 239. Graph of Dial Readings of a Superheterodyne Which Facilitates Tuning in Stations

Handy Logging Chart.—Each station tuned in should be carefully logged, and after a number of stations of different wavelengths have been logged, a chart of the kind shown in Fig. 239 can be made. A sheet of graph paper is obtained and the right side is marked off for wavelength readings from 200 to 600, and the lower edge is marked off in dial readings. **Dots** are made at the points where stations are logged, and when a number of station dots have been located, it will usually be found that they all lie on a curved line, the one of the oscillator condenser being just a little different from the one of the wavelength condenser. After such curves have been drawn, new stations can be found more easily as their wavelength is known, and one merely has to follow the curve until it crosses the horizontal line corresponding to the wavelength of the station. The vertical line crossing the curve near this point gives the dial reading.

Harmonics.—It is often found that the oscillator dial tunes in every station at two separate points. The point where the weaker signal comes in is called the "harmonic." The lower the wavelength of the signal which is tuned in, the less space there will be between the proper oscillation point and the harmonic point, and vice versa. On some superheterodyne circuits, harmonics have been eliminated, but most superheterodyne receivers have one harmonic and sometimes two. This is a normal condition and does not indicate trouble. It is a good idea to draw a separate curve on the logging chart for harmonics. If the receiver is located in the vicinity of broadcasting stations and it does not tune sharply, owing to low-frequency intermediate transformers, it is sometimes necessary to tune the oscillator dial to the harmonic instead of to the regular oscillation point, in order to bring in distant stations without interference from local stations.

CHAPTER VIII

REPRODUCER TROUBLES AND MAINTENANCE

The electrical impulses that are picked up by an aerial, and amplified by the tubes in a receiver, are transformed to sound waves by means of a reproducer, such as headphones or loudspeaker. The reproducer consists fundamentally of an electromagnet through which the output current of the receiver is fed, the fluctuations in the current causing similar fluctuations in the strength of the magnet, which attracts or repels a disk,

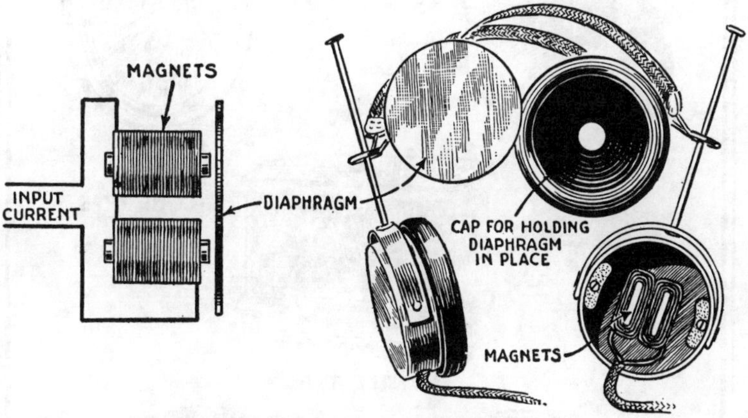

Fig. 240. Left: Magnets and Diaphragm of a Headphone. The Current Flowing Through the Magnets Energizes the Cores and Actuates the Diaphragm. Right: Headphones with the Cap and Diaphragm of One Earpiece Removed

as shown in Fig. 240. The attraction and repulsion of the disk is so rapid that it is really a vibration, producing sound. The rate of the vibration changes with the intensity of the current passing through the coils, which produces sounds of different pitch and tone quality. There are various types of loudspeakers as shown in Fig. 241, but they all operate fundamentally in the same way, whether the magnet or magnets act directly on the disk, whether on a movable armature, which in turn is connected to a small disk or a large cone by means of a connecting rod, or whether the movable armature is also a small magnet.

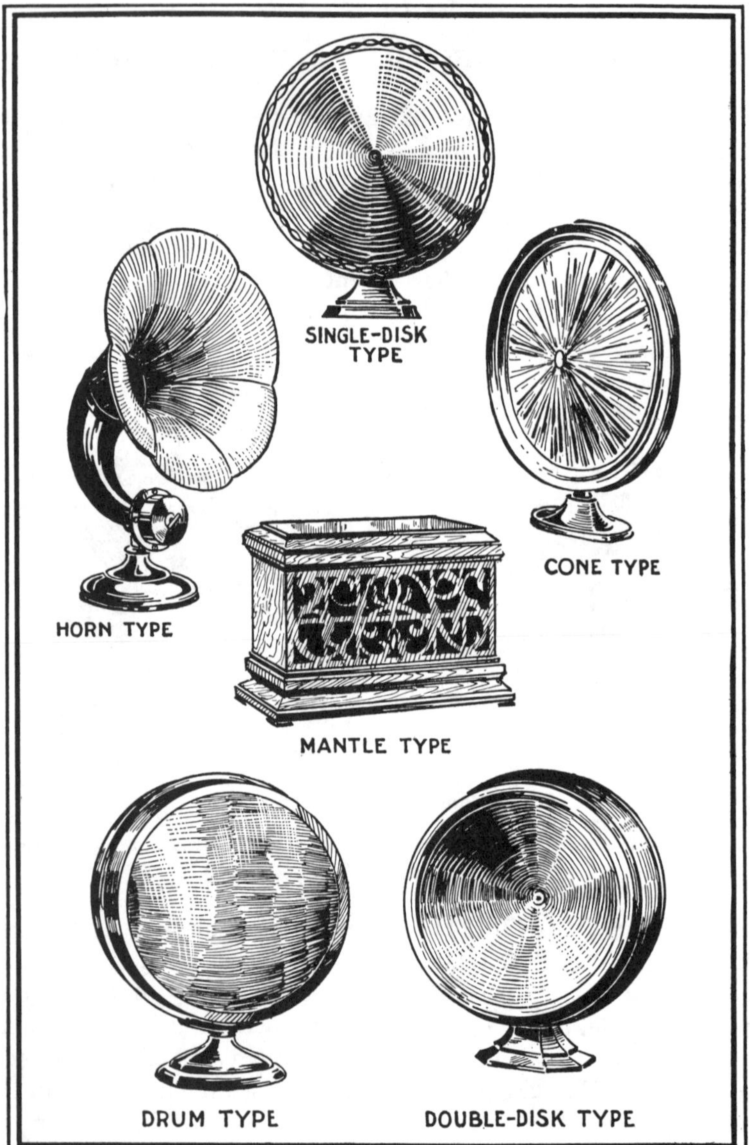

Fig. 241. Various Types of Loudspeakers. Besides Giving Faithful and Undistorted Reproduction of Sound Many of These Are Beautiful in External Design

Plug and Cord Troubles

Headphones and loudspeakers obtain their electrical energy over a flexible, two-wire cord, which permits the user to handle the headphones and to place the loudspeaker in any convenient position. The cord is fitted with small tips at both ends to facilitate making connection with the loudspeaker unit and with the receiver output terminals. Either binding posts or jacks are used as output terminals, and if the latter is used, a special plug has to be provided into which the cord tips are inserted, as shown in detail 1 of Fig. 242. The plug consists of a sleeve slipped over an insulating tube and washer, which in turn holds a rod.

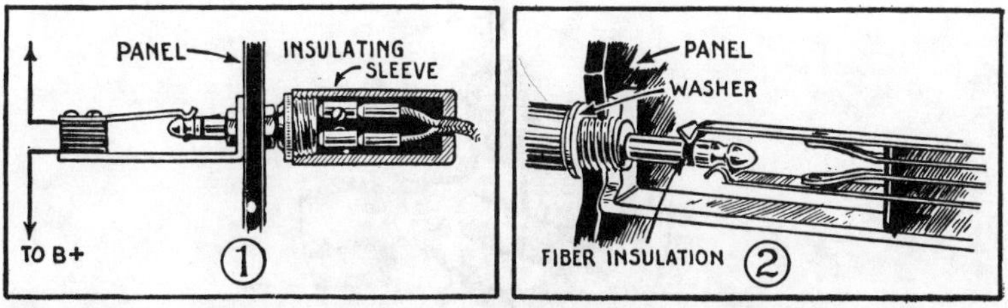

Fig. 242. (1) Cross-Section Showing how a Phone Plug Fits into a Jack and how Phone-Cord Tips Are Connected to the Plug. (2) Defective Plug Which Does not Fit in Jack Properly

If for any reason the sleeve and the rod make contact, the loudspeaker will be shunted out of the circuit entirely. This is sometimes the case when the insulating washer is faulty or has been accidentally omitted in the manufacture of the plug. Sometimes a plug is too short for the jack, causing the fiber insulation on the plug to come directly under one contact of the jack, as shown in Fig. 242, detail 2. The result is that the phones or loudspeaker connected to the plug fail to function. To remedy this trouble remove the washer behind the threaded bushing on the plug, or move the jack contact forward, which is possible in some cases by loosening a screw on the underside. Plugs sometimes develop loose contacts where they make connection with the cord tips. This is especially true of plugs using spring clips into which the cord tips are pushed to make contact; the clips may become bent, and will then not hold the tips securely. Better electrical contact is assured when the tips have to be pushed into sleeves, and small setscrews are used to hold them in place securely. However, the spring clips will be found handier to use.

Cords used for headphones and loudspeakers are usually made of two tinsel conductors covered with a heavy, woven cotton insulation. In the manufacture of cords the ends of the tinsel are doubled and wound tightly with a coil of fine wire and this is inserted into the tip, which contains melted solder. Tinsel is extremely flexible and is particularly adapted to this use, but it is very difficult to solder; hence the wire winding, to which the solder will readily adhere. Cords are apt to break as a result of excessive handling and bending, especially near the tips,

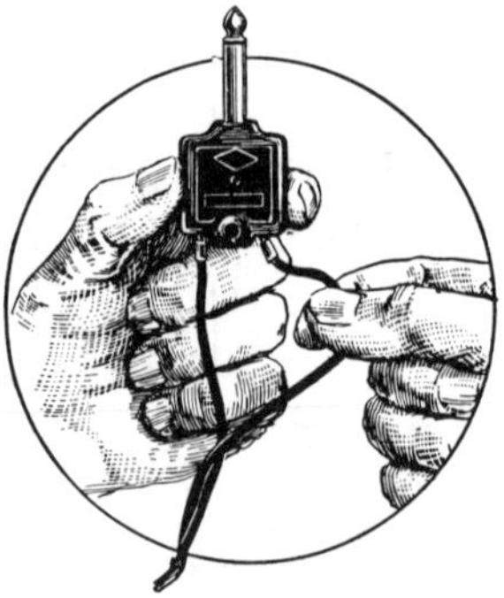

Fig. 243. Phone Cords often Break near the Tips Where They Are Handled Most Frequently

(Courtesy of "Popular Mechanics Magazine")

where most of the strain comes, as shown in Fig. 243. However, they can also be broken at other points, and if the storage battery is in an exposed position, some of the electrolyte may get on the cord, which will ruin it within a very short time.

In most cords one of the conductors has a striped insulation to distinguish it, and this cord is connected to the positive terminal of the phones or loudspeaker. The free end should be connected to the terminal on the jack or binding post, that carries the B-positive current, if no output transformer or impedance device is used. The body or frame of the jack, with which the sleeve part of the plug is connected, carries the B-positive current. Most plugs have markings indicating negative or positive polarity, which facilitates connecting cords to them correctly. If, however, such an indication is not given, and the plug is not of the separable type so that the owner can trace the internal connections, then the best method of determining the polarity is to plug in the loudspeaker and tune in a station. Turn down the rheostat until a moderate volume of reception is obtained. Reverse the connection of the cord tips in the plug. If better reception and slightly more volume is obtained, this is the correct connection. On many types of loudspeakers, especially of the cheaper grade, the polarity of the cord tips does not appreciably affect reception. If the loudspeaker is designed for definite polarity, the correct connection is necessary, and if the polarity is reversed, the loudspeaker unit will lose its efficiency.

Many cords have an extension of the woven insulation left

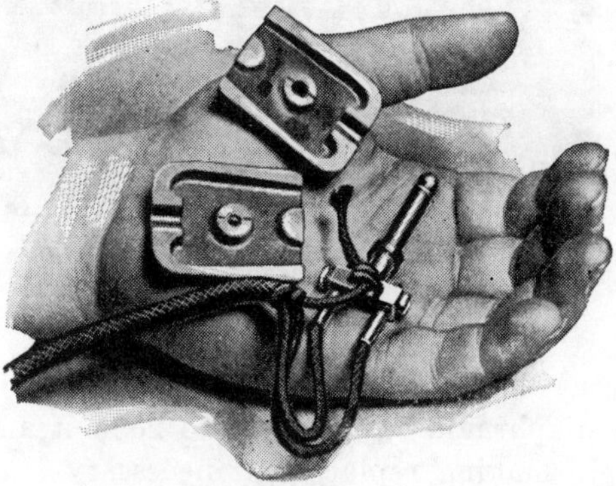

Fig. 244. Parts of a Plug Showing how the Extra Cord Is Tied to the Plug Body

(Courtesy of "Popular Mechanics Magazine")

at one end in the form of a string. This can be utilized to prevent undue strain on the conductors by merely tying the string to the plug, as shown in Fig. 244. When the cord is accidentally pulled, the strain is then taken by this string and by the plug itself.

Care of Headphones

Headphones are susceptible to more abuse than loudspeakers, as they are handled considerably. Dropping or jarring them tends to destroy the molecular arrangement in the magnet core, with the result that it may become slightly demagnetized. This makes an appreciable difference in their sensitivity. The screw cap on each phone must be kept tight, or rattling and distortion may result. Constant use often tends to loosen the caps, so that it is well to examine them periodically. If it is necessary to remove the cap and the diaphragm from the phone units, when installing a new cord, or for any other reason, the cap should be unscrewed carefully and the diaphragm slid off from

Fig. 245. When Removing the Diaphragm for the Phones Always Slide It off Carefully as Shown in the Illustration

the magnets, as shown in Fig. 245, instead of pulled off. Pulling the thin diaphragm off is likely to bend it and then it will not work well, making replacement necessary.

Loudspeaker Distortion

Good reproduction of sound depends on both the loudspeaker and the receiver. A good loudspeaker connected to a poor receiver, or a poor loudspeaker connected to a good receiver, usually results in an inferior quality of reception. However, strange as it may seem, a poor loudspeaker will often work better than a good loudspeaker on a poor receiver. This is because the poor loudspeaker does not reproduce all the "spurious" frequencies caused by the poor receiver, whereas a good loudspeaker, having a higher range, reproduces most of the spurious frequencies. This sometimes causes a rattle. As a rule, spurious frequencies are caused by overloading the amplifier tubes, and a rasping and blatting effect may also be evident. The best method of preventing such trouble, caused by overloaded tubes, is to substitute power tubes, which are capable of a greater output with less distortion.

A good loudspeaker should have a frequency range of from 50 to at least 5,000 vibrations per second, although there are sound waves having frequencies lower than 50 and higher than 5,000. There are very few loudspeakers that have a lower range than 200, and the majority of them operate most efficiently at 1,000 to 1,500 vibrations, having a tendency to become harsh and "tinny" when reproducing higher and lower frequencies. It should be remembered, however, that the same effect may be obtained on a really good loudspeaker if the audio transformers have a limited range.

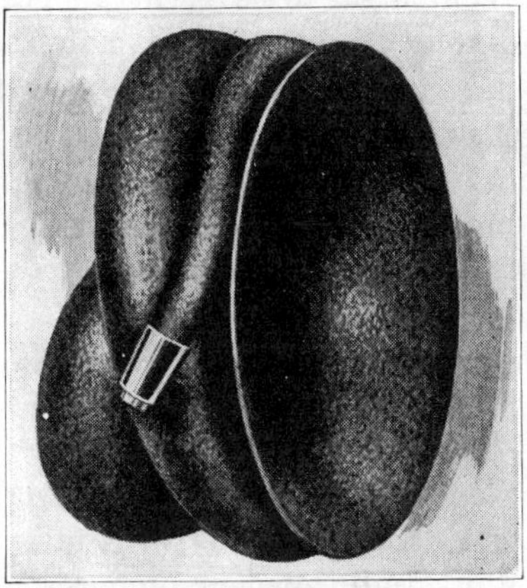

Fig. 246. Sound Chamber of Loudspeaker Having an Exceptionally Long Air Column

(Courtesy of "Popular Mechanics Magazine")

Distortion can also be caused by microphonic action of the detector and the first audio tube, as a result of vibration from the loudspeaker. The remedy is to remove the loudspeaker or to place it on a shock-absorbent rubber pad. The volume of a loudspeaker should be adjusted according to the size and the furnishings of the room in which it is located. Too much volume in a bare room of large dimensions will cause an echo, and give the listener the impression that the sound issuing from the loudspeaker is distorted.

Rattling may also be caused by hard particles lodging on the diaphragm, especially in case of the horn type of loudspeakers, in which the diaphragm is mounted horizontally. Loose internal and external parts of the unit may also produce rattling noises. Excessive sharpness of pitch can often be eliminated by shunting a small fixed condenser of from .001 to .01 mfd. capacity across the speaker terminals, in order to by-pass the higher frequencies. Apparently this seems to lower the tone, but the effect is really a softening and muting of the signal. The resonance of the sound chamber of horn-type loudspeakers may cause distortion, or it may have a damping effect on the signals. Sound chambers of metal and thin wood vibrate in resonance with the signals and produce a metallic or a "snare-drum" effect. The shape and taper of a horn from the throat to the bell and the length of the air column have as great an effect on its amplification value and tone quality as the material of which it is made. If the sound chamber has many turns or sharp corners there will be a reflection of sound, and signals will then be reproduced very unfaithfully. An efficient horn having an exceptionally long sound chamber, and having no sharp turns or corners is shown in Fig. 246.

Controlling Loudspeaker Volume

On many receivers having only one output connection, this being on the last audio stage, local reception may be so loud that it is uncomfortable, even if the rheostats are turned almost entirely down. Many radio owners then resort to detuning on the dials, but this practice does not give the full tone quality which is otherwise possible. A better method is to provide a variable resistance in the output line of the receiver. A handy form of variable resistance, which can be added to any receiver not having an output transformer or impedance device, and does not necessitate any wiring changes, is shown in Fig. 247. The resistance is incorporated in the loudspeaker plug, and it has a steady range of adjustment.

If it is desired to provide a resistance on the panel for this purpose, follow the diagram shown in Fig. 248. A variable resistor from 0 to 5-megohms resistance is used, and it is connected in series with the B-positive terminal of the pack or binding post. In some cases it may be found that better results can be obtained if the resistor is shunted across the two output leads instead of in series with one. This shunted arrangement is usually preferable if two speakers are connected in parallel.

Fig. 247. Modulator Plug, Consisting of a Variable Resistor and a Plug Combined, Serves to Control Loudspeaker Volume
(Courtesy of "Popular Mechanics Magazine")

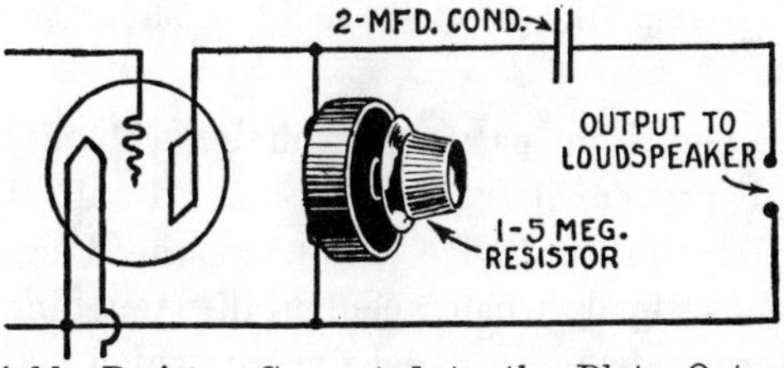

Fig. 248. Variable Resistor Connected to the Plate Output of the Last Audio Tube to Control Loudspeaker Volume

Eliminating B-Current from Loudspeaker Windings

When an ordinary UX 201A or similar type of tube is used in the last audio stage, it is usually not necessary to prevent the passage of direct current through the windings of the loudspeaker, but if a power tube such as a UX 171 or UX 210 is used in the last stage of audio, it is absolutely necessary to prevent the high-potential B-current from passing through the loudspeaker windings, or the speaker may be ruined. B-current can be kept from the windings by using an output impedance or transformer. An additional advantage of such an output device is that feed-back currents through the B-batteries or B-eliminator are suppressed. The method of connecting an impedance device in the circuit is shown in Fig. 249. The plate voltage of the power tube is fed to one terminal of the loudspeaker, while the other terminal of the loudspeaker is connected to one side of a 2 to 4-mfd. condenser. The other side of the condenser is connected to the maximum B-positive line, while the impedance device is connected to both plate output of the tube and to the B-positive line. In case of excessive volume a by-pass condenser of about .01 mfd. capacity is shunted across the leads connected to the loudspeaker terminals. This prevents the high-

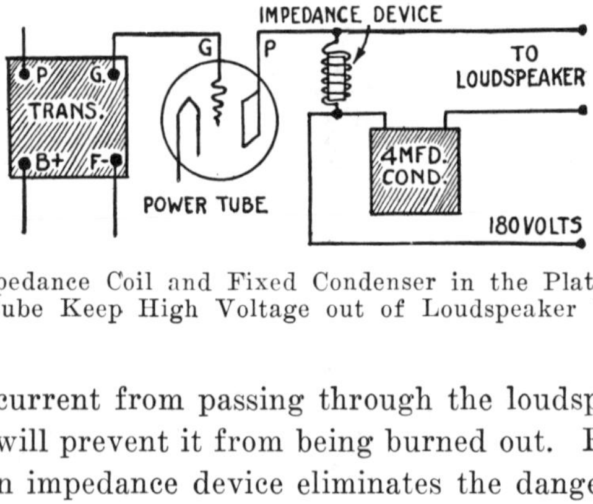

Fig. 249. Impedance Coil and Fixed Condenser in the Plate Circuit of a Power Tube Keep High Voltage out of Loudspeaker Windings

potential B-current from passing through the loudspeaker windings, which will prevent it from being burned out. Furthermore, the use of an impedance device eliminates the danger of anyone getting a serious shock when accidentally touching one of the loudspeaker terminals and a panel fixture, which may be charged with B-negative potential. Such indirect coupling of the loudspeaker to the receiver through an output device also helps to improve the tone quality of the receiver, prevents possible regeneration from this part of the set, and eliminates the frequent trouble of demagnetizing the loudspeaker magnets, by running the current through it in the opposite direction from that for which the unit was designed. An output transformer is usually connected as shown in Fig. 250, its primary being connected to the output of the receiver and its secondary to the loudspeaker cord.

Open-Circuited Loudspeaker

If it is impossible to obtain a click when the loudspeaker plug is inserted into the output jack, and one is reasonably certain that the receiver is in good condition, there may be an open circuit in the plug, cord, or in the loudspeaker winding. The best method of determining this is to plug in a pair of headphones. If signals cannot be heard, the trouble lies in the receiver, but if signals can be heard, the trouble lies in the loudspeaker end.

The exact point of the open circuit must then be determined. This is done by substituting another plug and another cord, and if there is still no response one can be sure that the loudspeaker

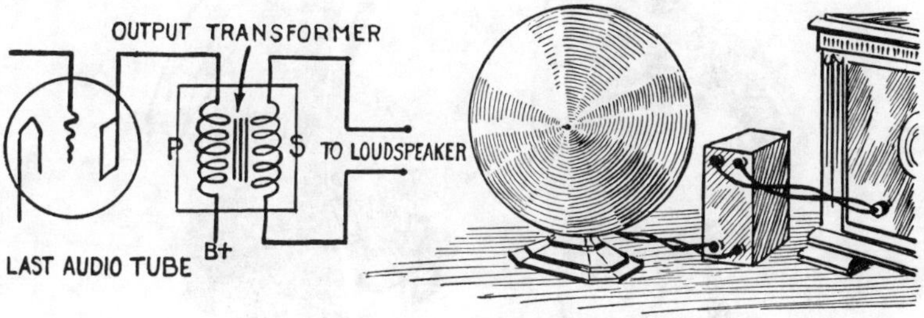

Fig. 250. Left: Wiring Diagram Showing how an Output Transformer Is Connected in the Plate Circuit of the Last Audio Tube. Right: Using an Output Transformer Outside of the Receiver

winding is at fault. As a rule, a break in the winding occurs near the end and can readily be repaired by taking the unit apart and locating the break in the coil. The wire used for magnet winding is very small and must be handled with considerable care as it will break easily. If the break is near the end of the winding, the end of the coil must be soldered to the terminal. The wire is insulated with a coat of enamel and this must first be cleaned off with very fine sandpaper, so that a good electrical connection can be made. As the soldering point is too large for small confined places, such as loudspeaker units, it will be found convenient to get a length of bare copper wire and wind it around the end of the iron, letting part of the wire project. With this the soldering can readily be accomplished.

Adjustment of Loudspeakers

Many loudspeaker units are provided with thumbnuts or other means whereby adjustments can be made. On horn-type loudspeakers there is often a knurled screw or wingnut on the unit, as shown in Fig. 251, by which variation can be obtained. Too tight a setting of this adjustment may cause rattling and too loose a setting a loss of sensitiveness. The proper method of making this adjustment is as follows: Plug in the loudspeaker and tune in a station. Turn the adjustment screw on the unit to the loose position, which is usually toward the left. Then

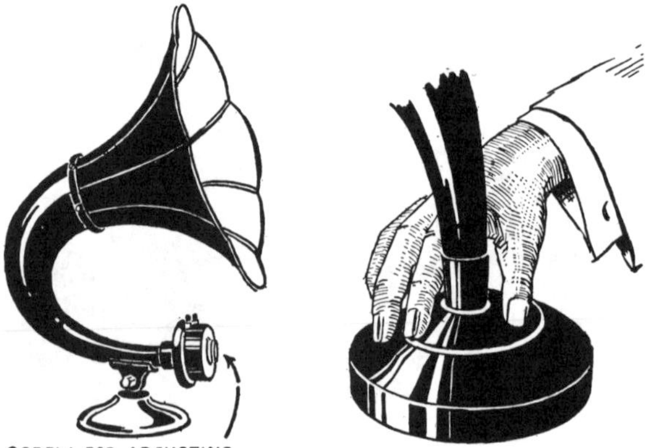

Fig. 251. Many Loudspeakers Are Provided with a Screw for Adjusting the Diaphragm in Order to Obtain Faithful and Distortionless Reproduction

screw it in gradually. It will be noticed that the volume will slowly increase up to a point where the signal will suddenly become muted, distorted, or rattling commences. This may or may not be accompanied by a click. The adjustment screw is again reversed slowly to the point where reception comes in best, and it is left at that setting.

The cone type of loudspeaker, which eliminates the need of a horn, has become extremely popular. A greater range of frequency, elimination of horn vibration and distortion, convenience, and better appearance, are the features claimed for this type. The unit and the cross-section of a cone-type speaker are shown in Fig. 252. A thin drive pin extends from the magnet armature through the center of the cone through two metal cone apexes, or through a metal collar attached to the disk, a setscrew being provided on the collar to maintain a firm connection between the cone and the drive pin, for the latter must actuate the whole cone. As the cone is subject to expansion and contraction, resulting from atmospheric changes, it is sometimes necessary to readjust the setscrew. All that is necessary is to loosen it and tighten it again. This adjustment must be made while the receiver is shut off. If it is attempted while the loudspeaker is in use, the adjustment will be incorrect. It sometimes happens that the drive pin is slightly bent, and bears with greater pressure against one side of the collar than the other side. This will cause an unequal distribution of force on the cone and consequently a distortion of

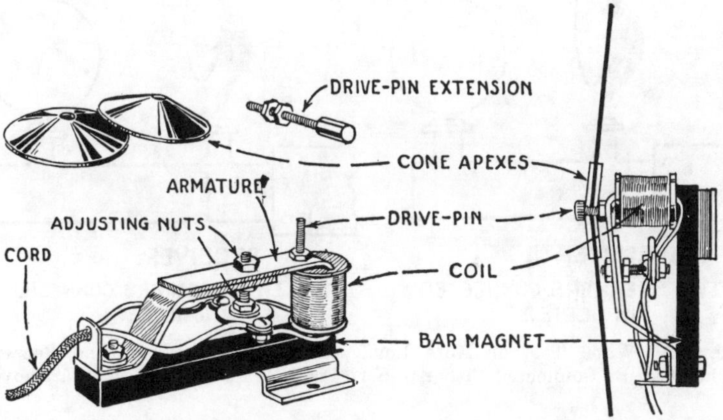

Fig. 252. Parts and Complete Assembly of a Cone Loudspeaker. The Details Show the Drive Pin, the Drive-Pin Extension, and the Cone Apexes

sound. The drive should be centered correctly in the collar so that it rattles when the setscrew is loose and the speaker is operated. The rod can be centered accurately by carefully adjusting the unit. In some cases it may be necessary to bend the rod a trifle. If the drive pin is threaded and provided with small nuts, as shown in Fig. 252, adjustment can be obtained by loosening these nuts so that the disk will not pull or push the drive pin, after which they are again tightened against the apexes. A cone speaker should not be set too close to other objects, for if it touches them a rattle will result. Similarly, care must be taken not to get any hard object inside of the cone.

Using Two Loudspeakers to Get Greater Frequency Range

Loudspeakers vary considerably in their frequency range. Some reproduce the higher notes more faithfully and are high-pitched, while others reproduce the lower notes, and are low-pitched. Obviously, the use of a high and a low-pitched speaker, connected so that they will operate simultaneously, will give a greater frequency range than is possible by using each one separately. There are two methods of connecting horns together, as shown in Fig. 253. The two speakers can be placed on oppoiste sides of a room, running the connecting wire over the

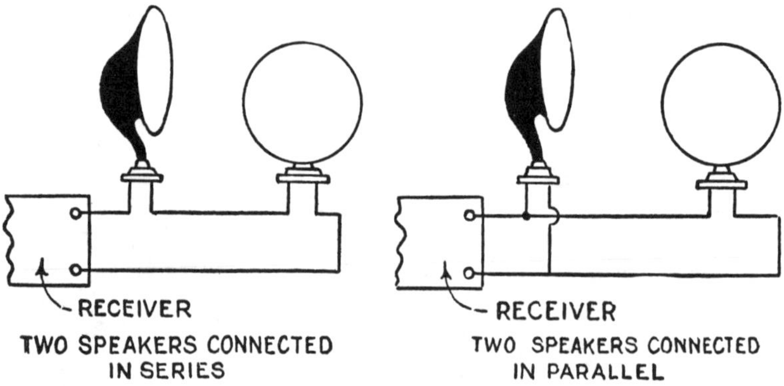

Fig. 253. When Two or More Loudspeakers Are Used on One Receiver They Can Be Connected Either in Series or in Parallel as Indicated Above

moulding. This arrangement distributes the sound more uniformly and will be found better than having them both placed on one side of the room.

Remote Loudspeaker Installation

Many radio owners wish to have a loudspeaker in more than one room, and this necessitates some interior wiring. Concealed wiring in the walls is advisable if the installation is to be a permanent one. Such an installation is shown in Fig. 254. When a house is being built, it is an easy matter to wire it for loudspeakers in various rooms, but if the wires have to be fished through the partitions, the task is much more difficult. The simplest method of doing the work is to make a small hole in the wall where a loudspeaker outlet is desired. The hole should be large enough to admit the passage of a length of ordinary furnace chain. A 5/8-in. hole is drilled in the sill of this partition, exactly under it and between the same two studs. The furnace chain is then inserted in the upper hole and the lower end of the chain is fished through the hole in the sill. A length of flexible metal-covered conduit, long enough to extend from the loudspeaker outlet to a similar outlet near the receiver is

Fig. 254. Houses or Apartments Can Readily Be Provided with Loudspeaker Outlets in Several Rooms. All the Outlets Are Connected in Parallel

(Courtesy of "Popular Mechanics Magazine")

attached to the chain, with which it can then be pulled up. A suitable box, a flush plug receptacle, and a brass plate to fit over it are installed at each loudspeaker outlet. This installation is similar to that of a concealed lead-in wire. The tips of the loudspeaker cord are securely fastened under the screws in the head of a plug, which fits in the receptacle. It will be found convenient to mark the polarity of the terminals so that the loudspeaker polarity will be correct. If a number of loudspeakers are used in this way, it is merely necessary to tune in any desired station at the receiver, and then plug in one or more loudspeakers. This makes it necessary to connect all the outlets in parallel.

Automatic Cut-Off for Loudspeaker When Using Phones

On some receivers installed in consoles no jack is provided on the panel for headphone reception. When it is desired to use a pair of headphones for distant reception, or when one does not wish to annoy others with loudspeaker reception, it is necessary to disconnect the loudspeaker cord from the output terminals, connect the headphone cord, and then reverse the same process when one has finished using the phones. As this entails

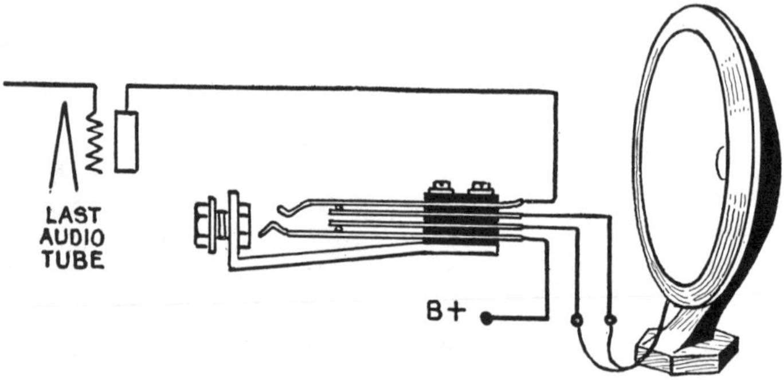

Fig. 255. Installation of a Jack Which Permits Using Headphones Where Loudspeaker Reception Is not Desired. When the Phones Are Plugged in, the Loudspeaker Is Automatically Disconnected

considerable trouble, it will be much better to install a jack on the panel, as shown in Fig. 255, provided the headphones are used often enough to make such an installation worth while. The headphone plug is inserted into the jack and this automatically disconnects the loudspeaker from the circuit. When the headphone plug is removed, the loudspeaker is again automatically connected. The jack used for this purpose is of the double-circuit type.

Multiple Switch for Comparing Loudspeakers

Dealers are often called upon to compare several types of loudspeakers for their customers. Instead of connecting and disconnecting each speaker separately, it will be found much more convenient to use a multiple switch of the kind shown in Fig. 256. It consists of a number of midget jacks, contact points, and a switch lever, arranged on a bakelite or hard-rubber panel, and fastened on a shallow box so that none of the wiring is exposed. One row of the midget jacks is connected

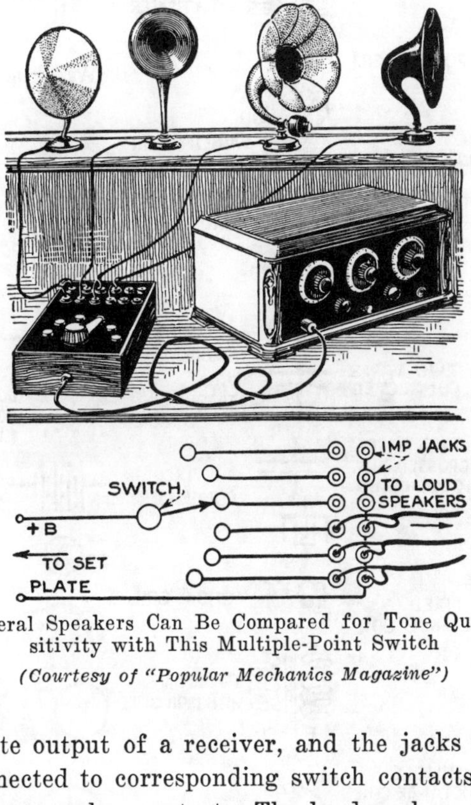

Fig. 256. Several Speakers Can Be Compared for Tone Quality and Sensitivity with This Multiple-Point Switch
(Courtesy of "Popular Mechanics Magazine")

with the plate output of a receiver, and the jacks of the other row are connected to corresponding switch contacts with which the switch lever makes contact. The loudspeakers are set on a shelf nearby, and connected to the switch as indicated.

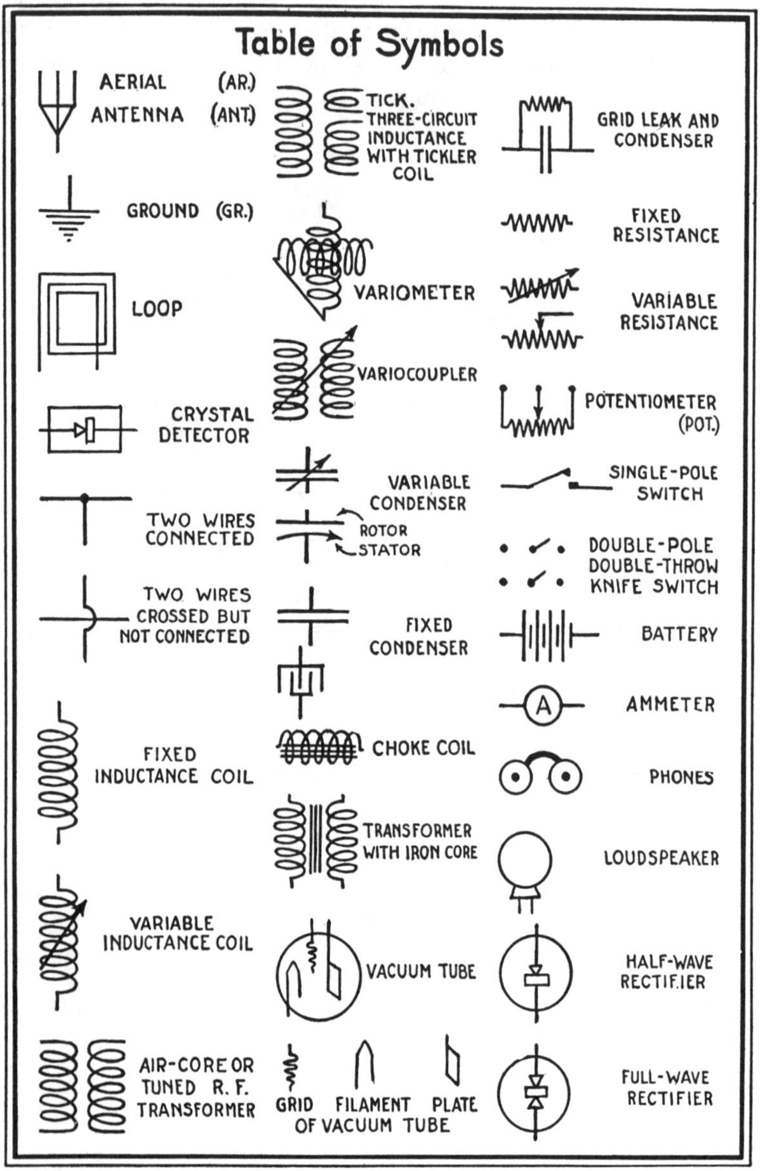

Fig. 257. Standard Radio Symbols with Which the Radio Owner Should Acquaint Himself in Order to Be Able to Read Schematic Diagrams

CONVENIENT TROUBLE-SHOOTING CHART

The following chart will be found to be of considerable value to the radio owner. It lists a great number of troubles that are often experienced, and also indicates the probable cause of these troubles. After determining from the chart where the trouble lies, one should, by using the index, refer to the section or sections in this book where detailed information is given on the particular trouble in question, and where the remedy is thoroughly explained.

TOTAL INAUDIBILITY

Symptom	Cause
Click heard but no signal comes in when loudspeaker is plugged in.	Aerial or lead-in open-circuited. Aerial grounded. Aerial and ground disconnected. Faulty and ineffective ground. Use of loop-designed receiver on outside aerial or vice versa. Primary and secondary windings of tuning coil short-circuited. Broken "pigtail" connection on variable inductance or condenser. Defects in radio-frequency stages. Short-circuited or open-circuited grid condenser. One or more internally open-circuited tubes. Improper contact of one or more tubes in sockets. Static blanket. Dead spot. Shielding effect of steel structure. Loose connection or high-resistance joint.
No click heard when loudspeaker is plugged in although all tubes light.	Worn-out B-batteries. Disconnected or open-circuited B-supply leads.

TOTAL INAUDIBILITY—*Continued*

Symptom	Cause
	Defective or poorly adjusted B-eliminator. Internally open-circuited B-battery. Breakdown in eliminator—usually a condenser or a resistor. Lack of A. C. supply for eliminator. Open-circuit in grid or plate leads. Defective transformers or resistors. Reversed C-battery. Break in loudspeaker cord. Defective loudspeaker plug or jack. Open-circuited winding of loudspeaker. Loudspeaker diaphragm adjusted too tight.
All tubes fail to light.	Discharged storage battery. Worn-out or broken-down storage battery. Discharged dry-cell A-battery. Defective A-eliminator. Disconnected or open-circuited A-supply leads. Too small wire used for A-supply leads causing high resistance. Burned-out tubes. Loose socket prongs not touching tube tips. Defective filament switch. Open-circuited filament wiring in receiver.
Some tubes fail to light.	Open-circuited rheostat or ballast. Some burned-out tubes. Loose socket prongs. Part of wiring open-circuited.

TOTAL INAUDIBILITY—*Continued*

Symptom	Cause
Inability to tune in distant stations.	Limited range of receiver. Atmospheric conditions. Aerial is too short. Use of a substitute aerial, loop or indoor aerial for regular outdoor aerial. Worn-out tubes which need reactivation. Low A and B-batteries. Careless tuning by operator.

WEAK RECEPTION

Symptom	Cause
Occasional weak signals.	Static conditions. Corroded tube tips and socket prongs.
Constant weak signals.	Receiver is located a considerable distance from broadcasting stations. Aerial is too short. Use of indoor aerial on receiver designed for outdoor aerial. Too loose adjustment of loudspeaker unit. One or more defective tubes. High-resistance joint.
Signal comes in louder when ground is disconnected.	Grounded aerial.
Rheostats must be turned up excessively high for fair volume.	Tubes worn out and need reactivation. Low A-battery.
Signal is strong when receiver is turned on but soon dies down.	Low A or B-battery, or both.

Weak Reception—*Continued*	
Symptom	Cause
Weak signals accompanied by slight whistling or crackling noises (may also oscillate).	Low B-batteries. Incorrect B-voltages.
Better volume obtained when tubes are changed.	Some tubes work better for r.f. and other for a.f. amplification.
Signal strength varies with the use of different phones and loudspeakers.	Demagnetized phone or loudspeaker unit. One loudspeaker better designed than another. Reversed phone-cord connections causing wrong polarity.
Tubes do not light up sufficiently and receiver has poor volume.	Corrosion on battery terminals causing high-resistance joint and preventing required amount of current from passing.
Weak signals after battery has been changed.	Reversed A-supply leads. Reversed C-supply leads.
Weak signals after lead-in and ground wires have been re-connected.	Lead-in and ground wires reversed on receiver binding posts.

Fluctuating Reception	
Symptom	Cause
Sporadic fading of signal strength.	Atmospheric conditions.
Intermittent fluctuations of reception.	Variations in B-supply from eliminator due to line surges.
Signal strength varies when hand is removed from tuning dial.	Body capacity.

Faulty Tuning	
Symptom	Cause
Broad tuning.	Proximity of receiver to broadcasting station. R. F. coils too closely coupled. Aerial is too long. Antenna effect of eliminator. Inadequate tuning circuit.
Too sharp tuning.	R. F. coils too loosely coupled. Aerial is too short for distant reception. Use of indoor aerial on receiver designed for outdoor aerial. High-peaked intermediate-frequency transformers of superheterodyne receivers.
Receiver breaks into oscillation easily when tuning signals carefully, causing an "oscillation hangover."	Too low value of grid leak.

Distorted Signals	
Symptom	Cause
Loud notes distorted.	Audio tubes overloaded. Poorly designed transformers.
Metallic reproduction.	Metal loudspeaker bell.
Constant distortion, especially on high and low notes.	Poorly designed and cheaply made transformers. Poor tubes having low mutual conductance. Intercoupling between non-shielded audio transformers. Too high transformer ratios. Incorrect value of defective resistors in resistance-coupled a.f. amplifier.

Distorted Signals—*Continued*	
Symptom	Cause
Distorted signals when r.f. and detector rheostats are turned up too high.	Oscillating r.f. tube or detector.
Distortion from certain broadcasting stations.	Poor broadcasting equipment and faulty operation.
Distortion when two stations on nearly the same wavelength are broadcasting at the same time.	Overlapping wavebands prevent tuning in either one sharply resulting in mutual interference.
Distortion on signals from distant stations.	Receiver tunes too sharp and cannot get benefit of the full waveband.
Harsh-toned reception.	Too high grid and plate voltages. Detector filament rheostat turned up too high. Incorrect resistors in resistance-coupled receiver. Too high grid-leak value.
"Fuzzy" reproduction.	Slow-leaking condenser in eliminator. Poorly adjusted eliminator. Incorrect value of grid leak.

Noisy Reception	
Symptom	Cause
Crashing and crackling noises at certain times.	Static discharges. Power lines grounded by trees, wet insulators, etc. Defective power transformer. Dirty commutators on electric motors causing sparking brushes. Passing street cars, interurbans, etc. Farm lighting unit.

NOISY RECEPTION—*Continued*

Symptom	Cause
	Short-circuited buffer condenser in eliminator. Arcing across anode leads of eliminator rectifier, across prongs of rectifier-tube socket, or across output of power transformer or eliminator. Defective eliminator resistors.
Buzzing sounds.	High-frequency electrical apparatus. Rotary spark gap of transmitting station.
Rasping, spluttering or crackling noises when receiver is being tuned.	Loose connection. Thread-like burrs on plate of variable condensers.
Constant crashing and crackling noises (sometimes evidenced with weak signal strength).	Noisy B-battery, especially when old and nearly worn out.
Sporadic hissing, crackling and frying noises.	Internal short circuits in tubes.
Howl, increasing gradually in pitch and volume.	Microphonic detector or other tube (caused by too high plate voltage or by vibration).
Ringing noise when receiver or cabinet is tapped.	Microphonic detector.
Steady humming sound.	Aerial parallel to electric power lines. Lead-in or ground wire, loop, or receiver in close proximity to extension cords or electric lamps. Poor filtration of eliminator. Inductive effect from eliminator.

NOISY RECEPTION—*Continued*

Symptom	Cause
	Defective buffer condenser of eliminator. Trickle charger in operation while receiver is used.
"Motorboating" (putt-putt sounds).	High internal impedance of filter section of eliminator. Faulty resistors and leady condensers in eliminator. Fluctuations of voltage of a. c. supplying eliminator. Excessive voltage on r.f. and detector tubes. Incorrect value of resistor in detector-plate circuit of resistance-coupled receiver. Incorrect grid bias. Defective grid condenser. Too high grid-leak resistance. Reversed winding of a.f. transformer. Unmatched impedance between power tube and loudspeaker.
Squawking noises when tuning.	Set turned up too high for proper operation.
Howling and whistling noises.	Too much regeneration. Pick-up from other reradiating receivers. Too high grid leak. Unneutralized r.f. stages. Inductive effects between parallel grid and plate lines. Inductive effect of a loop. Two stations at slightly different wavelengths operating simultaneously. Intermediate stages of superheterodyne oscillating. Incorrect value or defective filter condenser in superheterodyne.
Clicking while tuning.	Oscillating r.f. tubes.

INDEX

INDEX

C

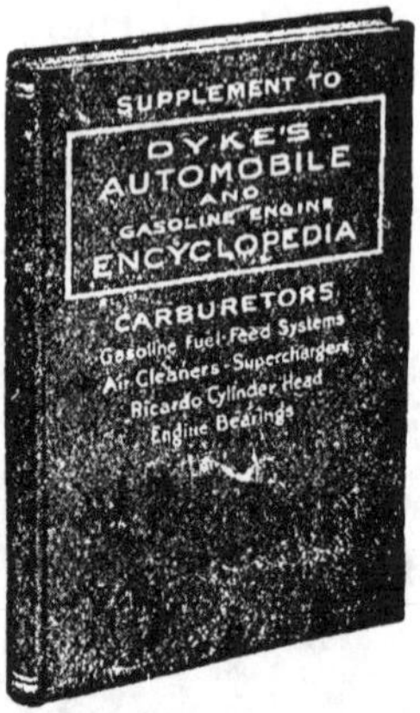

CONDENSED TABLE OF CONTENTS OF SUPPLEMENT